UCF
46.00

The Triplet Genetic Code
Key to Living Organisms

The Triplet Genetic Code
Key to Living Organisms

Lynn E.H. Trainor
University of Toronto

World Scientific
Singapore • New Jersey • London • Hong Kong

Published by
World Scientific Publishing Co. Pte. Ltd.
P O Box 128, Farrer Road, Singapore 912805
USA office: Suite 1B, 1060 Main Street, River Edge, NJ 07661
UK office: 57 Shelton Street, Covent Garden, London WC2H 9HE

British Library Cataloguing-in-Publication Data
A catalogue record for this book is available from the British Library.

THE TRIPLET GENETIC CODE
Key to Living Organisms

Copyright © 2001 by World Scientific Publishing Co. Pte. Ltd.

All rights reserved. This book, or parts thereof, may not be reproduced in any form or by any means, electronic or mechanical, including photocopying, recording or any information storage and retrieval system now known or to be invented, without written permission from the Publisher.

For photocopying of material in this volume, please pay a copying fee through the Copyright Clearance Center, Inc., 222 Rosewood Drive, Danvers, MA 01923, USA. In this case permission to photocopy is not required from the publisher.

ISBN 981-02-4467-3
ISBN 981-02-4468-1 (pbk)

Printed in Singapore.

This book is dedicated to the students who studied Physical Theory in Biology, and particularly to the graduate students who helped me establish the strong connection which exists between theoretical physics and the understanding of biological phenomena. These graduate students were:

Jeff Beis, Wendy Brandts, Al Busch, Frank Jones, Dave Jourrard, Gary Knight, Louis Lemieux, Charles Lumsden, Kathy Nolan, Glenn Rowe, Jeff Sutton, Victor Szabo, Patrick Tevlin, John Totafurno, Charles Trainor and France Tremblay.

Along with their harvest of scholarships, these students were effectively scientific pioneers, treading biological pathways where most theoretical physicists deign not tread.

Foreword

> Your piercing eye
> will dim and darken; and death will arrive,
> dear warrior, to sweep you away.
> - *Beowulf*: 1766-68.[1]

As Book 6 of Homer's *Iliad* opens, the war between Greece and Troy, having dragged on for ten long years, rages with new fury. Bronze-clad warriors clash on the windswept plain before the Trojan city. Blood and mangled bodies are everywhere. Greek heroes, larger than life, drive the Trojan forces before them in what looks like a rout until, drawing near their city gates, Troy's beleaguered defenders suddenly gain new hope. Turning to face the invaders, they grasp their weapons with fresh strength. Shields are raised but, arrow-quick, Homer whips away from the massed carnage to a close-up. Just two men, a Greek and a Trojan, step forward to confront each other on a patch of bloodied plain laid waste by the war. Diomedes, much-feared champion of the Greek forces and confident of his prowess, is in no hurry to deliver his Trojan adversary the final blow. Instead he savors a lingering speech, relishing his confidence to outdo all mortal men and challenging the Trojan Glaucus to make his ancestry known. Is he a death-bound human and so ready to meet his fate? Or, perchance, one of Olympus' deathless gods, unfit for Diomedes' sword?

Glaucus is not hugely intimidated by this show of Achaean bravado. As the battle swirls round them he squares on Diomedes and, in one of those Homeric similes resonant across three millennia says to him, famously,

hoiê per phullôn geneê toiê de kai andrôn[2]

[1.] The translation from the Anglo-Saxon is Seamus Heaney's. See his luminous *Beowulf: A New Verse Translation*. Bilingual Edition. New York: Farrar, Straus and Giroux, 2000.

[2.] This is *Iliad* 6:146. See David B. Monro and Thomas W. Allen. *Homeri Opera in Five Volumes*. Oxford: Oxford University Press, 1920. You will find the anglicized (Roman

The generations of leaves as they come and go across the seasons, the generations of humankind as they come and go across the years: they are but alike. Concerns about ancestry, in the end, matter little. Whether Greek battle lord or Trojan retainer we stand not apart from the Nature's flickering transitions, but within them. We do not escape. Glaucus then shares some family history. (In fact, the two discover their families have been friends, and so whomever else they may slay that day the two warriors must be allies.)

Today, almost three thousand years after Homer, we are adrift on the same river of time that swept Diomedes and Glaucus together on the plains of Troy. But this is about to change, dramatically and permanently. We are the witnesses to, and perhaps the largely reluctant participants in, an unprecedented moment in human existence, indeed of all life on Earth. For the first time since life arose on our watery planet three billion years ago, a species (us) has discovered, and is rapidly mastering, the means to reshape and remake its biological essence. That means every particle of the human organism, from genetic molecule through to finished being is now up for grabs. As a collective humanity, it is just possible that we will pause to reflect on what this means, for ourselves and the planet we steward; more likely, with the unknown calling and fortunes to be made, we will just plunge right in. That will be it. The generations of leaves may thereafter still come and go as they always have done across the flicker of seasons, but our human generations will break away. They will start to plot their own course, at their own tempo, shaping human body, brain, mind, and lifespan in directions we can as yet hardly conceive. To what end?

Put crudely, we got to this turning point in two steps, both of which you will explore in Lynn Trainor's elegant new account. In step one, we cracked a code. The hereditary molecule DNA was at last understood to be a linear text, a sequence of four molecular letters that somehow spells out the protein fabric of living cells and helps to shape where and when this fabric spins itself into finished creatures. The letters are grouped into a code—a genetic code with deep chemical and mathematical properties that, at a stroke, illuminate the nature of inheritance and the origins of life.

---

alphabet) Greek text version of the *Iliad*, along with two fine English translations, instantly to hand at Tufts University's *Project Perseus,* http://www.perseus.tufts.edu The (very) free contemporary Canadian rendering of 6:146 immediately following the Greek line is my own.

Foreword

That was step one, and it revolutionized the life sciences during the 1950's 60's and 70's. In step two, beginning chastely in 1980's and with an accelerating rush over the past decade, we are, literally, reading the DNA text of many organisms from start to finish. Armed with our knowledge of the genetic code, we are beginning to tease out the relationships whereby the DNA text shapes body and mind. This is the Human Genome Project, because it includes (some would argue has as its centrepiece) the complete DNA text of human beings. The complete human DNA text has, in the past few months, been worked out, at least in first-draft form. Up till now, over the vast stretch of time from the first, primitive living cell to our hominid ancestors and they to us, the composition of the DNA text, with its consequences for the form of our bodies and minds, has been shaped by the blind forces of evolution. From this point on, given knowledge of the genome, we ourselves are stepping in to replace evolution's blind outcomes with our own, willed goals for our bodies and minds. Who will decide what these should be?

There is enough beauty, enough adventure, and enough science in the genetic code and its discovery to fill any account. We are fortunate, however, that Lynn does not stop there. Instead, he moves firmly from matters of scientific fact to those of theoretical understanding and then to the compelling ethical issues that we, as a conscious, creative, genome-empowered species, must now, having cracked the code, deal with. It's the intellectual equivalent of a dip in a mountain stream: bracing, for the author is no starry-eyed futurist. He belongs instead to what I call the "Canada gang"—a diverse, restless group of Canadian thinkers like Harold Innis, Marshall McLuhan, Derrick de Kerckhove, Ursula Franklin, John Ralston Saul, Thomas Homer-Dixon, David Suzuki, Charles Taylor, and myself who, left rather cold by millennial visions of manifest destiny, at-all-cost globalization, and utopia through technology, are inclined to ask tough questions about what technological change does to people, their nations, and their societies, and who gets to decide what changes are good.

From time to time we hear the argument that matters of science and society and our future are hard stuff, best left to people with (supposedly) big brains and plenty of fancy degrees (or least lots of cash and dot.coms). That is rubbish: too much is at stake for anyone to be uninformed about DNA and what it means. The Universe, despite its abiding mystery, can be an astonishingly simple place, provided you have an able guide to its core truths. Lynn Trainor's skills and credentials as a guide to this hidden world are masterful. The decisions about our biological future we must make, as individuals and as a society, rest on our understanding of the DNA story, whether our debates zero in on genetically altered foods, genetic testing of the unborn, personal privacy, or the genetic engineering of the human species. In preparing ourselves, and in the sheer pleasure of insight, *The Triplet Genetic Code* sets a lead we can all follow, a step in a lifetime's reflection on what, at last, we are each seeking in the dream to be human.

Charles J. Lumsden

Preface

In a democratic society, it is essential to have an informed public on the issues of the day concerning social philosophies, life styles, ethical considerations and decision making. Increasingly these issues have their origins in the applications of new scientific knowledge arising out of study and research. Among the new challenges linked directly to scientific activity, perhaps no other is as profound and revolutionary for society as that of molecular biology. The purpose of this book is to bring an understanding of the genetic code and its significance to a wider audience than the professionals in biochemistry and genetics, but including professionals in other fields who would like a little more than an acquaintance with this field, but who do not have the time nor circumstances to devote themselves to an in-depth and detailed study of molecular biology.

A particular audience for which this book should be appropriate are the students and teachers in secondary schools and community colleges for whom a concise account of the nature and impact of molecular biology in the modern world is an essential part of a wider curriculum in the biological sciences.

Biochemistry constitutes a vast field of knowledge coming from scientific discovery. Its methodology is fraught with details on how to carry out canny experiments on micro-organisms and molecular events, but with a history of profound discovery on the self-organization of living systems, how their structures are achieved and their functions carried out at the molecular level. Ultimately, this detail is essential to scientific discovery. But this book is purposely restricted to the results of these discoveries—the grand emergence, so to speak, of the vast collective effort of biochemists, biologists, geneticists, mathematicians, physicists and chemists, who have contributed to the experimental techniques and physical concepts leading to explication of the genetic code.

The present text has leaned heavily upon several detailed, professional studies on various aspects of molecular biology. Primary among these have been the following:

(1) The Scientific American collection of readings, *The Molecular Basis of Life,* edited by Robert Haynes and Phillip Hanawalt [40][3]. This collection contains essays written by several of the prominent scientists who themselves contributed in a major way to the discovery and elucidation of the genetic code.
(2) The remarkable and imposing text by Bruce Alberts *et al.* entitled *The Molecular Biology of the Cell* [2] and its update [3].
(3) M.V. Volkenstein's excellent book *Molecular Biophysics* [92].
(4) The text *Biochemistry* by L. Stryer [87] with its attractive colored illustrations and its clarity of style.
(5) The impressive and informal behind the scenes account of the discovery and development of the genetic code in *The Eighth Day of Creation* by H.F. Judson [52].
(6) The informative update of Mae-Wan Ho's book *The Rainbow and the Worm* [42] with its focus on some challenging problems for modern biology.
(7) The book *Physical Theory in Biology* (editors C. J. Lumsden, W.A. Brandts and L.E.H. Trainor) [59].

While molecular biology promises great success in achieving a profound understanding of biological processes with applications to modern technology and human welfare, it also has a dark side potentially inimical to the future stability and welfare of the biological world, in general, and to the welfare of humanity, in particular. The balance between exciting promises and potential disasters will be difficult to achieve, but it is essential in a democratic society that we make every effort to do so. Our best efforts in achieving this balance will require an educated and prudent public. Our very being demands it. The matters involved are too significant to be left entirely to the scientific professionals and to the vagaries and unregulated driving forces of the market place.

[3.] In this book, numbers in square brackets [#] refer to references while numbers in curly brackets (#) refer to glossary items.

Acknowledgements

I am indebted to the many people who have helped me put this book together. I am especially indebted to Professor Charles Lumsden of the Department of Medicine at the University of Toronto for his design of the book cover and dust jacket, also for his generous help in bringing to my attention many valuable reference materials and for much help with final editing. I am also indebted to him and to Professor Ellie Larsen of the Zoology Department for reading an early draft of the manuscript and for their many suggestions and constructive criticisms for improving it. The generous assistance of Ming and Jennifer Tam with this project was a great treasure. I am particularly indebted to Ming for formatting the manuscript and preparing it for submission and in camera-ready form for the publisher. This work involved drafting of several diagrams. The expert assistance of Mr. Khader Khan and Mr. Raul Cunha of the Physics Department in preparing several diagrams and in the final drafting and digital printout of the book cover and dust jacket is also gratefully acknowledged. I wish also to express thanks to Barbara Chu in the Physics Library for valuable help with the references, and Anne Trainor for assistance with final editing.

I wish to express sincere appreciation to my son Charles, to my patient friend Dr. Ronald Jhu, "Mr. Fixit," and to my colleague of many years, Professor Sam Wong of the Physics Department at the University of Toronto, for their patient and valuable assistance in keeping my word-processing system alive both here in Toronto and in my study on Saltspring Island on the beautiful West Coast of British Columbia.

Finally, I wish also to express my gratitude to Alan Pui of Singapore and Stanley Wu-Wei Liu of Rivers Edge, New Jersey for their encouragement and guidance.

Copyright Permission for Figures

Copyright permissions to redraw or adapt the figures in this book are gratefully acknowledged:

Academic Press, London, UK
- Figure 13, from Rowe, G.W. "A Three-Dimensional Representation for Base Composition of Protein Coding DNA Sequences," *Journal of Theoretical Biology*, 1985, **112**, page 433.
- Figure 14, from Trainor, L.E.H., Rowe, G.W. and Szabo, V.L. "A Tetrahedral Representation of Poly-Codon Sequences and a Possible Origin of Codon Degeneracy," *Journal of Theoretical Biology*, 1984, **108**, page 459.

Joan Starwood, Elburton Georgia, USA
- Figure 8, adapted from Ms. Starwood's diagram on page 30, article by Robert Haynes in the Scientific American Reprint *The Molecular Basis of Life*, R. Haynes and P. Hannawalt (editors). San Francisco: W.H. Freeman, 1968).

Springer-Verlag, New York, USA
- Figure 12, from Rowe, G.W., Szabo, V.L. and Trainor, L.E.H. "Cluster Analysis of Genes in Codon Space," *Journal of Molecular Evolution*, 1984, **20**, page 172.

World Scientific, Singapore (via Academic Press, London)
- Figures 15, 16, 17 and 18, from Figures 6, 7, 8 and 9 in Trainor, L.E.H., Rowe, G.W. and Nelson, G.J. *Physical Theory in Biology: Foundations and Explanations*. Lumsden, C.J., Brandts, W.A. and Trainor, L.E.H. (editors). Singapore: World Scientific, 1997, pages 405-429.

Reproductions by permission of Routledge, Inc. part of The Taylor and Francis Group as follows:
From Alberts, B., Bray, D, Lewis, J., Raff, M., Roberts, K. and Watson, J.D. *Molecular Biology of the Cell*, 3rd edition, 1994, New York: Garland Publishing:
- Figures 2 and 3, adapted from Panel H, page 60.
- Figure 5, adapted from Figure 6, page 101.

Adaptations from Stryer, L. *Biochemistry*. San Francisco: W.H. Freeman and Company, 1975:
- Figure 4, from Figures 24-9 and 24-10.
- Figure 7, from Figure 27-4.
- Figure 9, from Figure 2-17.

Contents

FOREWORD vii

PREFACE xi

ACKNOWLEDGEMENTS xiii

COPYRIGHT PERMISSION FOR FIGURES xiv

1. INTRODUCTION 1

2. WHAT IS LIFE? 7

3. THE NUCLEIC ACIDS, DNA AND RNA 11

4. PROTEINS — THE MOLECULES OF LIFE 25

5. PINNING DOWN THE CODE 31

6. DESCRIPTION OF THE GENETIC TRIPLET CODE 39

7. ORIGIN AND DEVELOPMENT OF THE GENETIC CODE 45

8. A PHYSICAL APPROACH TO GENETIC ORIGINS 51

   8.1 Codon Bias in Viral Genes — A Thermodynamic Theory 52

   8.2 Cluster Analysis in Codon Space 56

   8.3 The Tetrahedral Representation of Codon Space 62

9. REDUCTIONISM VERSUS HOLISM 71

10. CULTURAL AND MATERIAL IMPACTS OF MOLECULAR BIOLOGY 77

GLOSSARY 87

APPENDIX A. PEPTIDE BONDS AND PROTEIN STRUCTURE 101

APPENDIX B. PROCEDURE FOR INTRODUCING A METRIC FREE DISTRIBUTION OF POINTS IN CODON SPACE 105

APPENDIX C. FUN WITH TETRAHEDRA 107

APPENDIX D. GENETIC ENGINEERING 109

REFERENCES 113

INDEX 119

The Triplet Genetic Code
Key to Living Organisms

1. Introduction

Theologians speak of revelation, the disclosure of knowledge by a supernatural agency. But science has its own revelations, e.g. Newton's concept of universal gravitation (1), that every piece of matter in the universe is being drawn by a force toward every other piece of matter in the universe. Its state of motion is determined by the resultant of all these forces. This concept accounts at once for the motions of the planets about the sun, for the daily tides in the oceans and seas, and for the formation of the stars from self-gravitating "dust clouds" out in the vast reaches of the cosmos. Another example from physics is the revelation of Maxwell (2) that all light is a form of vibrating electromagnetic fields propagating through space with the frequency of vibration accounting for the colors in the visible spectrum and other distinctive properties. In some minds, the revelations of Einstein (3) are the greatest, that space and time are relative concepts, and that all matter is energy in the sense that it is interconvertible with all other forms of energy: heat, light, stored energy, energy of movement, etc. These revelations came about as a result of observations on physical phenomena coupled with flashes of intuition arising from deep insight and informed speculation on the nature of our world. In each instance, the revelation is primarily associated with a single scientist (here, Maxwell in one instance, Einstein in the other), although the revelation represented a conceptual convergence contributed to by others. We use the term "revelation" because deep intuitions of this type reveal profound aspects of the physical world; nevertheless, intuitive processes are not well understood. The example of quantum mechanics is rather different, in that its discovery cannot be primarily associated with any single scientist; rather, it has a history of progressive piecewise development with many contributors, and many partial revelations.

Discovery of the triplet genetic code provides a history somewhat similar to that of quantum mechanics. Its revelation was also piecewise, with a history of discovery spanning years and involving many scientists contributing pieces to the puzzle. As with quantum mechanics, it is no less than a scientific revelation of sweeping proportions, with profound ongoing implications not only for the biological sciences but also for philosophy, the physical and social sciences, economics, and particularly, ethics.

The purpose of this book is to bring to the interested reader an appreciation and an understanding of what the genetic code is, and why it has come to revolutionize thinking about living systems as a whole, particularly as regards the connection between structure and function in those systems. Perhaps this is best accomplished by telling the story of how molecular biology came about, in particular the identification of DNA (4) as the "genetic stuff," and the deduction of the essential structure of DNA by Watson and Crick [95] and finally, after much effort, the determination of the genetic code itself—which informs the cells as to which strings of amino acids will constitute the proteins necessary for structure and function in whatever living organism. ("Function," as used here, signifies the activities in any organism necessary to its survival, including its regulatory processes.)

However, a word of caution about historical exercises must be entered here. From our present perspective, we can now identify important milestones which marked the paths to discovery, but these are not necessarily the perspectives available to the researchers when these milestones were achieved. The true history of discovery is probably impossible to reconstruct. In that sense our history will be but a sketch from our present vantage point. It will not and cannot do justice to the many research projects which contributed to major revelations along the paths of discovery. The details of these research projects necessary to their full comprehension and appreciation are too vast for our purposes, and would place demands on the reader inconsistent with our modest account of the key elements informing our knowledge of molecular biology.

In successive chapters we develop the story of what is called *The Central Dogma of Molecular Biology*, viz. that the flow of information is from DNA to RNA to protein.

We begin in Chapter 2 by discussing some important aspects of the nature of life:
- The great diversity of biological organisms [99], and the necessity of devising classification schemes, in the spirit of Linnaeus (5) for plant forms and Buffon (6) for animal forms, extending from the great plant and animal kingdoms down through the various subclassifications and finally to distinct species.

Introduction 3

- The similar molecular composition of organic forms in general, with species diversity of form and function accounted for in the organizational detail.
- The necessity for a biological theory to encompass both development and evolution (development or ontogeny having to do with how an individual organism, under direction from its "genetic stuff," develops and grows into a mature and functioning entity; evolution or phylogeny having to do with how species are formed, and how they change over many generations primarily as a result of chance mutations and the competition for survival (Darwinism (7)), resulting in successful adaptations to changing environmental circumstances).

Chapter 3 will briefly deal with the elucidation of the structure of the nucleic acids as the "genetic stuff" in the chromosomes by Watson and Crick [95]. This momentous discovery culminated from the work of many scientists who contributed to the identification of the nucleic acids as the genetic material and the delineation of some of their structural and chemical properties. They were particularly aided by the complementary work of Rosalind Franklin and R.G. Gosling [26, 27, 28]] on the one hand, and Maurice Wilkins and his associates on the other (see [98]) on diffraction analysis of the "DNA fibres" pulled from aqueous solution and later from their detailed X-ray analysis of crystalline DNA. It was this cooperative effort between the X-ray analysts (Wilkins, Franklin and their various associates) (8) and the model building (stereochemistry) analysis of Watson and Crick and their associates that lead to the elucidation of the double helical structure of DNA [97]. In Chapter 3, the nucleic acids will be described, and the ground established for understanding why these molecules have the properties which enable them to house the genetic code, and to make it available to direct the molecular processes which assure stable inheritance and the proper functioning of the myriad complex processes taking place in the cells of any given organism.

In Chapter 4 we will discuss the importance of proteins in organic systems, and survey their pervasive occurrence, e.g., as primary structural elements in cells and cellular systems, and as specialized agents (e.g., enzymes, hormones, neurotransmitters and antibodies) for driving or regulating a multitude of complex biological activities.

We first discuss the structure of polypeptide molecules, in general. These are polymers consisting of strings of amino acids joined together by peptide bonds (9), discussed in Appendix A, before specializing to the important subclass of proteins. Members of the protein subclass of biological interest are all constituted from a small pool of just twenty amino acids (10). What makes proteins special in biology is their property of folding up in an aqueous environment to form highly specific 3-dimensional structures, which can play specialized and exotic roles in organisms. The specificity of each protein in its form and function depends entirely upon the sequence of amino acids in its polypeptide string. One can identify several structural levels—the primary structure is, of course, the succession of amino acids themselves in the protein string, with the secondary and tertiary structures having to do with the details of the folding process leading to highly specific molecular structures (called conformations), their biological properties being determined largely by their sizes, shapes and electrical charge distributions.

We will then introduce the concept of receptor sites (11) and explain their roles in biological theory as the basis of our understanding of the efficacy and specificity of most biological functions, including the action of enzymes, the regulatory power of hormones, and the complexity of antibody activity in immune reactions.

In Chapter 5, we will return to the structure of the nucleic acids DNA and RNA and take up the story of how the genetic code was deduced following the revolutionary discovery by Watson and Crick, and how their model provided a molecular basis both for inheritance and for the processes leading to the development of individual biological organisms.

Chapter 6 will be concerned with a description of the genetic code itself—what it is as a language and how it is utilized in the transcription (12) process to transfer information stored in the DNA to mRNA (messenger RNA), which carries it to the translation (13) sites, where it is used to direct protein synthesis in the special ribosomal (14) parts of the cell.

Chapter 7 will discuss theories of the origin and development of the genetic code itself from prebiotic times to the present era, in particular, how the almost universal nature of the genetic code brings an incredible unity to the biological world at the basic level of information storage and utilization.

The discussion in Chapter 8 will specialize to the basic question of: why a triplet code? Since the code specifies the sequence of amino acids on any particular protein and all proteins in biological organisms are built from a common pool of only twenty different amino acids, why was a triplet code selected in nature? Further, how has nature dealt with the technical problem of using the 64 possible codons (15) in the triplet code on DNA to specify uniquely and accurately the necessary sequences of amino acids (from the pool of 20) for building any required protein in any of the plethora of possible organic systems?

In Chapter 9, we try to develop a perspective on the nature of the revolution in molecular biology and genetic determinism. In particular, we consider possible limitations to knowledge and understanding arising from the extreme reductionist viewpoint characteristic of this revolution. We further consider some alternative approaches inherent in such holistic concepts as nonlinear dynamics, complexity theory, chaos theory and quantum statistics. Of particular interest from a theoretical point of view is the question of how the coherence of biological activity is achieved in multicellular systems, a problem discussed by Mae Wan Ho [42].

Finally, in Chapter 10 we look to the future of molecular biology to discuss some of its implications for human activity, particularly its effects on various ethical considerations in a complex and changing world.

2. What is Life?

The mystery of the code is directly tied to the mystery of life itself. Every living organism experiences life, by definition. Every conscious person senses what life is; however, attempts to give it precise definition fail in one way or another—this is primarily due to the great diversity and complexity of biological forms (see *The Diversity of Life* by Edward O. Wilson [99]). But whatever definition one might wish to adopt, trying to understand and deal with the pervasive occurrence and diversity of life forms is a daunting challenge. From an analytical point of view, organisms are the most diverse and complex systems known; but despite their great diversity and complexity, one perceives a remarkable degree of commonality among them. To quote from page 3 of Haynes and Hanawalt [40] "life is a collective word that subsumes a vast panorama of complex phenomena associated with all the organisms that exist ... One is relieved of this impossible assignment [of understanding all life forms each in its own terms] by the prospect that there are certain basic characteristics to be found in the simplest creatures and yet are essential for even the most complex."

One attempt to categorize life and indicate its basic chemical and physical basis was the classic book of Schroedinger, *What is Life?* [81] which inspired generations of scientists, Jim Watson [94, p.18], in particular. Schroedinger set the tone for inquiring what is life really about, in physical and chemical terms. The bottom line of Schroedinger's argument was that one should be able to eventually understand life in these terms without the need for vitalism. A modern sequel by Mae-Wan Ho [42] complements Schroedinger's book in two ways: first, by giving an update, and second by giving particular emphasis to the remarkable unity and coherence of biological organisms.

Despite great diversity in shapes, sizes, and appearances, organisms do not differ substantially in their basic molecular composition. The diversity is in the detail of molecular arrangements. They all utilize the same categories of macromolecules—the nucleic acids DNA and/or RNA, the hydrocarbon lipids (16) involved in cell membranes, the polysaccharides (17), the ubiquitous proteins, etc. Moreover, all organisms depend on outside sources of energy to drive the various mechanisms essential to the lives they fulfill. All organisms are composed of one or more living cells. The study of cellular structure hence gives focus to much of biological investigation and analysis. Modern molecular biology has come a long way in providing answers to many of the questions which naturally arise in dealing with these complex systems, such as how the stability of inheritance is achieved in the face of the vast array of complex chemical and physical processes taking place within each cell during its life cycle. Moreover, how do individual cells in a multicellular system come to "know" what their specific functions are (i.e., how do they achieve differentiation (18)), and how do they carry out these functions consistently and reliably? As in physics, theoretical biology deals with these problems by identifying general principles having wide applications throughout the biological world. The discovery of the remarkable structures and properties of DNA and of RNA cleared the way to finding answers to these and other questions. In this remarkable story, the genetic code plays a central role, as we shall see.

Although we are primarily concerned with the code itself—what it is and how it controls many aspects of the operation of the cells of an organism—it is essential to first understand the structures of the genetic macromolecules DNA and RNA which carry and transmit the code and make the whole business functional in the first place. Codes must be housed in the biological structures themselves, and mechanisms in the cells must exist for them to be accessed in an appropriate manner and utilized at appropriate times. The structure and properties of the nucleic acids are remarkably suited to playing a central role in making all of this possible. This fact will become more clear in the next chapter when we deal with the basic structures of these macromolecules.

Of particular interest, perhaps, in the theory of biological systems is their self-organizing properties (61). While the general principle holds in the cosmos that entropy (19) is increasing, i.e., that the world on the whole is becoming more disorderly, for a growing biological system its own entropy is actually decreasing as its general organization develops, which might seem a contradiction to the principle of entropy increase (second law of thermodynamics (20)). Indeed, it was a question of major concern among scientists until it was eventually realized that while the entropy of the self-organizing biological system was, in fact, decreasing, it was doing so at the expense of its environment. If one takes the environment into account, the overall entropy of system plus environment is increasing so that there is no contradiction with the second law of thermodynamics. Nonetheless one has, in self-organizing systems, this quality of a system of separating itself from its environment in such a way that the system can decrease its entropy at the expense of the environment. Detailed study of these processes shows that such a separation requires an external energy source to drive the system away from equilibrium, as well as a supply of the necessary metabolites for life. From this point of view, a biological organism is a system which has an input of "high quality" energy (i.e. useful energy), part of which is stored within the system in the process of self-organization, while the rest is discharged in some form of "low quality" energy (primarily as heat). Here "quality" is a measure of the extent to which the energy can be utilized to drive useful processes, such as chemical reactions, charge separations, mechanical work, etc. Heat energy by this definition is a low quality form of energy since it represents the kinetic energy of random molecular motions which cannot be harnessed entirely, and then only with the presence of a cold reservoir (21) into which the unused heat can be dumped. In other words, a biological organism, whatever else, is a self-organizing system which develops on a throughput of energy and metabolites, with a discharge of the unusable part of the energy and the unused or transformed parts of the metabolic input.

As outlined in the book *The Rainbow and the Worm* [42], biological processes are extremely efficient in the energy transfers vital to life. The primary source of energy, of course, is sunlight; this is high quality energy which must be transformed into a variety of forms in the living organism, such as energy of movement (including fluid flow), chemical energies for driving endothermic reactions vital to the cells, etc.

Another feature of living systems is their strategy of producing the proteins that they need upon ingestion. First the input proteins are broken down (catabolism) into their amino acid units and then reconstituted with the assistance of complex biochemical processes (anabolism) into the amino acid strings required for the proteins necessary for life in the particular organism involved.

The structure of proteins as distinctive strings of amino acids linked together through peptide bonds will be discussed in Chapter 4 and in Appendix A.

3. The Nucleic Acids, DNA and RNA

The elucidation of the structure of DNA by James D. Watson and Francis Crick in 1953 [95, 96, 97] depended in part on the exacting X-ray studies (22) carried out, somewhat independently, by Rosalind Franklin and Maurice Wilkins and their associates in the laboratory of J.T. Randall (8) at King's College, University of London. Initially, Wilkins observed that DNA fibres pulled from a DNA gel seemed to show ordered structure and Gosling in the same lab was able to obtain remarkable X-ray diffraction photographs of them [98]. Later when it became evident that DNA could be crystallized, the X-ray analysis could be extended. Eventually this work confirmed the double helical structure of DNA posited by Watson and Crick from X-ray photographs. A critical piece of the puzzle came from Rosalind Franklin's recognition of the B (wet) structure of DNA which greatly improved the X-ray pictures, and her insistence that the "backbone" (two helical chains) was on the outside rather than on the inside of the molecule. Watson and Crick, in a stunning *tour de force*, using molecular modelling techniques developed by Linus Pauling (23) and others for studying protein structures, were able to identify the principal details of the DNA structure, i.e., the identification of its chemical parts and how they are assembled. Although much work had still to be done to verify their proposed structure and confirm its relative universality, the essential step had been taken.

Among the difficult questions to be worked out was the central question how the DNA/RNA system worked in practice—what were the mechanisms that enabled this system to perform its remarkable functions [96]? Although the code was still unknown, what immediately appealed to Watson and Crick [95] was that the structure was suggestive of how DNA duplicates when cells undergo division so that both daughter cells have a full complement of the parent DNA. Heading the list of revealed problems was the determination of how inheritance is achieved at a molecular level and how the information contained in this genetic material was stored, read out and interpreted by the cell's machinery, providing a molecular approach for physiology. Progress in this latter regard required one to understand the specific nature of the code itself and of its utilization.

How, in fact, is genetic information actually stored in the DNA, and how is it brought to bear on the actual cellular mechanism required to activate and regulate the appropriate processes in the cell? The cellular processes sustaining life involve proteins in an essential way, as will be discussed in some detail in Chapter 4.

Cell dynamics is, after all, exquisitely complicated. Structures have to be built according to strict specifications in order to achieve essential functions and these processes must occur in the right time sequences. In short, the code has to prescribe what proteins must be produced by the cellular machinery, in what quantities, and in what time sequences. The information highway within the cell was gradually established as joining DNA to RNA to protein (the so-called Central Dogma of Molecular Biology). The process of reverse transcription in which DNA is constructed from an RNA template is referred to as reverse transcription. Thus, while the Central Dogma provides a good guiding principle, it is not universal. DNA stores the genetic information, mRNA copies it piecewise (24), specific strings of messenger mRNA corresponding to specific genes (with their protein products). The mRNA then carries the instructions to the cellular machinery (ribosomes) where the correct sequence of amino acids is assembled to make a protein required by the organism. The success of this procedure requires a great deal of regulation at various points and at various levels along the molecular pathways from DNA to protein.

Our simple picture does not, for example, explain how regulation of the protein production takes place. However, in a remarkable series of laboratory studies, Jacob and Monod (25) in France showed in an experiment in a metabolic context that the genetic complement of an organism includes "regulatory genes" which are dedicated to the regulation and control of molecular processes in the cell, as part of the general master plan of cellular activity. The Jacob and Monod model [48] is referred to as the *lac operon* model because it explains the regulation of lactose metabolism in a cell according to whether glucose is or is not available for metabolism. The *operon* concept involves a complex of structural and regulatory genes, the former coding for the protein enzyme, the latter for turning its production on or off. It was later found to apply to many circumstances involving enzyme regulation, particularly in cellular development and differentiation.

The term "gene" is a unit of genetic determination (first proposed by Gregor Mendel in 1863 [62] from patterns of inheritance in breeding experiments before the present era of molecular biochemistry). The gene is usually now identified with a section of DNA coding for a protein.

The structure of DNA as presently conceived, can be best understood by visualizing a molecular ladder, with two sides to which rungs are attached, which is then twisted into a double helical spiral with successive turns running along the length of the ladder, the two sides of the ladder then constituting the double helical structure with parameters indicated by Rosalind Franklin's X-ray photographs. The ladder now has the form of a spiral staircase (Fig. 1).

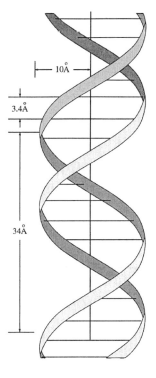

Figure 1. A schematic view of a section of the DNA molecule shown as a twisted ladder. The two helical ribbons represent the two side chains each consisting of a succession of sugar molecules (deoxyribose) strung together by phosphate linkages. The horizontal bars represent the Watson-Crick pairs which form the rungs of the ladder.

The essential structure and function of this DNA molecule is best appreciated by identifying a constant part of the DNA molecule, namely the two sides (helical chains) of the ladder, and a variable part, namely the succession of "rungs" binding the two chains together. By a constant part, we mean that the sides of the ladder correspond to long, relatively stiff strings or chains of repeating identical molecular units, sufficiently complex in themselves, nevertheless units which do not change as one proceeds along the spiral ladder from unit to unit. The chains consist of sugar molecules (deoxyribose) bonded together by phosphate links (26). Also attached to each sugar molecule and sticking out from the chain at right angles is one of the four bases that contribute to the rungs bridging between chains. The molecular unit consisting of the sugar molecule and its attached base is referred to as a *nucleoside*, which is shown schematically in Fig. 2.

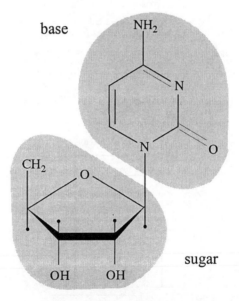

Figure 2. Depiction of nucleoside. A nucleoside is a molecular complex consisting of a sugar molecule with a base attached in such a way that the base sticks out at right angles to the chain direction. Pairs of bases, one from each side chain, bridge together to form the ladder "rungs."

When one considers the molecular complex of sugar, plus base, plus phosphate linkage it is referred to as a *nucleotide*, as illustrated in Fig. 3.

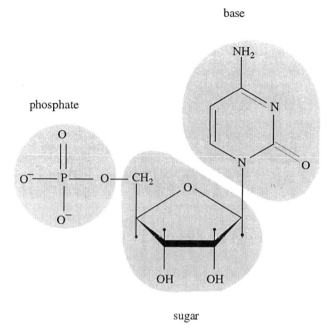

Figure 3. A molecular complex consisting of a nucleoside (Fig. 2) with its phosphate linkage attached is referred to as a *nucleotide*.

The "rungs" on the ladder are composed of two bases (27), one purine and one pyrimidine, bound together by hydrogen bonds as indicated in Fig. 4. Each base in the pair is bonded to a sugar molecule, one base to a sugar molecule on one chain, the other base to a corresponding sugar on the other chain so as to form a molecular bridge or "rung" between the chains, thus binding them together. Bonding of bases to the sugar molecules is such that the base molecules lie in planes perpendicular to the directions of the side chains, much in the nature of conventional ladder rungs. The purines and pyrimidines are rather flat molecules because of their carbon ring structures. These rungs form the "variable" part of the DNA molecule, because successive rungs can be of several types, as we shall see, so sequences of rungs can be distinguished from one another thus providing a mechanism for storing information much as is done with the digital structure of a computer disc.

The nucleosides are bonded together by phosphate linkages on each side chain to form the two long, rather stiff chains which form the sides of the ladder. The bases (27) attached to each sugar molecule stick out at right angles to the chain direction, the complex of sugar molecule, phosphate link and base forming a nucleotide (Fig. 3). Pairs of bases, contributed by the two nucleotides on corresponding sides of the ladder, bond together so as to form rungs which hold the two sides together. It is at once evident that information must be stored in this variable part (rungs) rather than in the side chains (constant part).

Now, it turns out that there are four possible bases (27) involved in the structure of the rungs of the DNA ladder, namely, thymine, adenine, cytosine and guanine. These bases are designated in the literature as T, A, C and G, respectively. A pair of bases, one contributed from each of the two side chains, bond together through hydrogen bonds (28) to form a relatively flat rung on the ladder.

The four possible bases divide into two groups: the purines and the pyrimidines. The purines each have two carbon rings in their structure, each ring containing two nitrogen atoms, while the pyrimidines consist of a single carbon ring, also containing two nitrogen atoms. Their chemical structures are such that purines and pyrimidines bond in pairs, one of each to a pair. Two purines do not link stably, nor do two pyrimidines link stably in the DNA structure (primarily because two pyrimidines are too small to bridge between the side chains, while two purine molecules are too large). Now comes the clincher. It turns out that for chemical bonding reasons, thymine always links with adenine through two hydrogen bonds (28), whereas cytosine always links with guanine through three hydrogen bonds (Fig. 4).

This means that one has basically only two types of rung on the ladder, rungs formed by AT bridging and rungs formed by CG bridging. However, since each strand of DNA forming a side of the ladder is directional (i.e. proceeding in one direction along the molecule is not the same as its reverse for reasons of molecular structure and the two strands run in opposite directions), one has to distinguish AT bridging from TA bridging, and similarly GC bridging from CG bridging. Thus each rung on the ladder can be any one of four kinds: AT, TA, CG and GC. The structure of DNA is indicated in Fig. 5.

Generally RNA differs from DNA [72] in that a fifth base, uracil (designated U), replaces thymine in bridging with A. Also, RNA uses a different sugar in its backbone (ribose rather than deoxyribose). As a result of these changes, the RNA backbone has generally only one sugar-phosphate

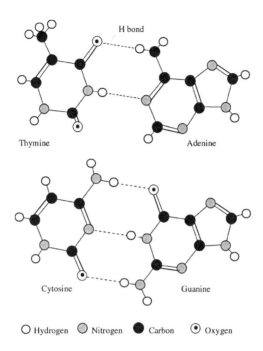

Figure 4. Each rung of the ladder corresponds to a pair of bases (one purine and one pyrimidine) bonded together by hydrogen bonds, as shown. Adenine bonds with Thymine (AT pair) through two hydrogen bonds, while Cytosine bonds with Guanine (CG pair) through three hydrogen bonds as shown. Figures 4 and 5 indicate the basic structure of the DNA molecule. The hydrogen bonds are indicated by dashed lines.

strand as indicated schematically in Fig. 6. However, under some conditions the RNA strand can take on a single helical conformation.

Returning to DNA, we see that the successive rungs are *variable* (that is they can vary in structure from one rung to another), and it turns out that this is how the information is stored in the DNA. As a result of this pairing structure of DNA, which we shall refer to as *Watson-Crick pairing* (29), the two nucleotide chains (sides of the ladder) are complementary structures—the knowledge of the sequence of bases on one helical chain of the ladder immediately predicts what the complementary sequence is on the other helical chain [16]. (It is significant that Erwin Chargaff [12] had observed what has become known as *Chargaff's rules*, namely, that when a sample of

DNA is chemically analyzed from whatever source available there are always equal amounts of cytosine and guanine, and again equal amounts of thymine and adenine in the residues, an indication that these pairings somehow must exist and play a role in the associated genetic molecules.) This observation by Chargaff turned out to be another important clue used by Watson and Crick in their deduction of the structure of DNA.

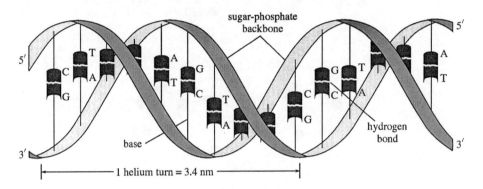

Figure 5. The rungs of the DNA helical ladder consist of base pairs, one contributed by each side chain, and bonded together in the manner indicated in Fig. 4. In the diagram the base pairs are separated by a white gap, the gap indicating hydrogen bonding of the various pairs of bases.

But where does the code come in? Why is this double helical ladder capable of holding the genetic code? It was several years after the discovery of the essential structure of DNA by Watson and Crick, that the code itself was uncovered. We now know that the code is carried by the succession of bases on either side of the helical ladder. From the complementary (Watson-Crick) pairing (29), if one side of the ladder has a sequence of bases GTCCATG, for example, the other side of the ladder has the complementary sequence CAGGTAC.

Of course the DNA molecule is very long (has many base pairs), even for simple organisms. The bacterium *Escherichia coli* DNA has about four million base pairs compared to human DNA which has upwards of three billion base pairs (see [55, p. 20]). The sleuthing job that had to be done was to discover just what is the code, assuming that the structure of DNA carries the code as then seemed evident. In other words, how was information stored in the DNA and how was it interpreted to specify the succession of amino acids (10) in any given protein? The answer to this question was neither immediate nor evident. Obtaining it represents a remarkable period of experimentation in molecular biology with many outstanding scientists and scientific groups contributing to the process of determining and deciphering the code. Notable among these were the laboratories of Marshall Nirenberg and Francis Crick (30).

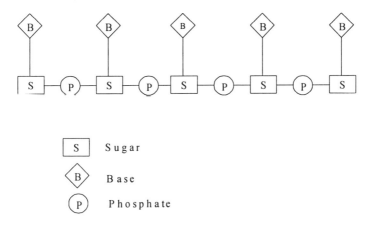

Figure 6. A schematic view of a section of the RNA molecule as a single chain. It differs from DNA in that the base thymine T is replaced by the base uracil, denoted U, and the sugar ribose replaces 2-deoxyribose. As a result the RNA chain is single-stranded.

When a nucleated cell divides, the daughter cells each carry a full DNA complement. This is achieved in the cellular division by a combination of enzymes, *helicases*, to unwind the parental DNA in a region, exposing single strands, and DNA polymerase which assists two new DNA chains to be polymerized (31) (assuming the necessary nucleotide materials are available in the nuclear cytoplasm), each one complementary to a strand of the original chain. The hydrogen bonds broken by DNA polymerase pair up with new nucleotides floating freely in the nuclear fluid, again according to the pairing rules A with T and C with G. Thus, two new DNA chains are produced with appropriate bases attached to duplicate the original DNA molecule. It turns out that this process, called *replication* (32) is "conservative," in that the two DNA molecules resulting from replication each have one chain from the parent molecule and one newly synthesized chain [63]. Thus in each cell division which characterizes tissue growth and repair, the DNA is duplicated so that each daughter cell has the full DNA complement of the parent cell.

The structure of DNA lends itself to this replicative process. But DNA has a second fundamental role to play in specifying the various phases of the cell cycle, i.e. prescribing what proteins must be produced to ensure the structure and carry out the necessary functions of each cell in the organism.

In the process of identifying the code, scientists also had to work out what cellular machinery is involved in the production of proteins, i.e., how *is* the information transcribed from the DNA by the mRNA molecules transmitted to the appropriate molecular machinery? And, in fact, what *is* the molecular machinery and how does it come up with the appropriate proteins specified in the genetic code? The detail is very involved and will not be pursued here; however, a general outline of what happens is set out below.

First, let us imagine the process which must be carried out to construct a protein specified by DNA. In the final stage amino acids, one at a time and in appropriate sequence, must be linked together by peptide bonds to form a particular protein chain. The information as to what is the appropriate sequence of amino acids required to construct a particular protein is contained in the DNA in coded form. (As we shall discuss in Chapter 5, it turns out that each succession of three bases, called a codon, in the correct "reading frame" (33), specifies a particular amino acid from the pool of twenty. Three codon exceptions are "stop" signals, i.e. they are not coded for at the ribosome and the protein assembly is terminated at that point. The stop codons are sometimes referred to as nonsense codons.) Moreover, although initiation of a gene sequence is more complicated than termination, it does involve certain initiation factors (special proteins) at the ribosome, as well as the sequence AUG on the messenger RNA feeding into the ribosome.

In transcription, the two DNA side chains are pulled apart in a region by a large enzyme, RNA polymerase, thus enabling an mRNA molecule (a single chain) to form through base pairing with an exposed DNA chain over the separated region. The polymerization of the appropriate ribose nucleotide units (selected by base pairing) from the nuclear fluid is catalyzed by the RNA enzyme. As the enzyme, RNA polymerase, moves along the DNA ladder it separates the DNA chains ahead of it (each region corresponding to a gene), while the DNA chains restore their previous pairing behind the polymerase, gradually releasing the RNA messenger molecule to separate off. A special region called the *promoter* on the DNA allows the RNA polymerase to begin and specifies its direction of travel along the DNA chain. A *stop* sequence of bases on the DNA chain also specifies when the RNA strand is completed. Thus the information for a complete gene contained in the DNA is transcribed into RNA. To complicate matters, in the construction of a gene for eukaryotic systems the transcribed RNA initially contains expressed regions, called exons, interrupted by unexpressed regions called introns which must be excised to form the final mRNA chain. ("Expressed" in this context means they carry information which will be used in the ribosomes to construct protein.) This excision is carried out by special enzymes. (The purpose of the introns in the initial transcript remains an enigma to this day, but it is usually assumed that they play some higher order role that is not yet understood.) Each mRNA molecule then has to pass through channels in the two-membrane sheath surrounding the nucleus and attach itself to an available ribosome in the extranuclear cytoplasm. The ribosomes, rather complex structures, are organelles (34) in the cytoplasm structured from a combination of proteins and ribosomal RNA (35) (another "structural" form of RNA designated rRNA).

What happens next is complicated, but suffice it to say here that the mRNA molecule with its string of codons threads its way through a part of the ribosome, one codon at a time. As it does so, special "adaptor" molecules (36) in the cytoplasm become attached successively to a particular site (active site) on the ribosome. These adaptor molecules, a particular form of RNA called transfer RNA, or tRNA for brevity, play a crucial role in matching the codon information at the active site on the mRNA chain to the information necessary for stringing together the correct sequence of amino acids for the protein under construction. (Note that by convention the codons are referred to sequences on messenger RNA, which by Watson-Crick pairing means that the template sequencing from the DNA strand to the RNA strand was formed from anticodons (37) on the DNA. Again, recognition of a codon on mRNA by an adaptor molecule (i.e. tRNA) requires that the tRNA molecule possess an anticodon matching region.)

Transfer RNA is a special sequence of bases which fold into a hairpin structure as shown schematically in Fig. 7. The molecules of tRNA are highly specific in that they have several modified bases (indicated by blank spaces in the diagram and by a ψ base in several places) and a special recognition site (anticodon) consisting of 3 bases (on the lower turn in Fig. 7) which pairs with the codon (37) being translated on the mRNA. (An additional complication in Fig. 7 is the appearance of a different base, inosine, which has a weaker but less specific base pairing, in the anticodon region. This feature is connected with Crick's *wobble hypothesis* and will not be discussed further here. It is discussed in detail in [87, p. 650].) The

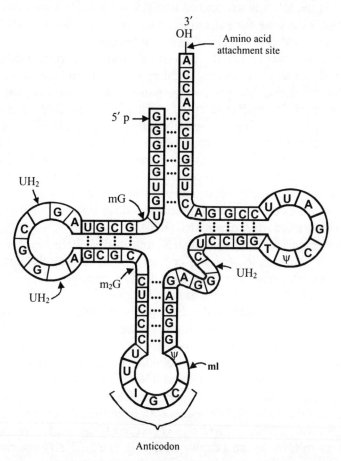

Figure 7. The general structure of the transfer RNA molecule. The molecular shape is due to a special sequence of bases which fold into a hairpin structure as shown. The molecules of tRNA are highly specific in that they have several (indicated by blank spaces) modified bases and a base sequence which acts as a recognition site for the crucial anticodon base triplet (lower loop) which is destined to pair with the appropriate codon on the mRNA, in this example a codon for alanine. (Adopted from Stryer [87, P. 645], with permission.)

appropriate amino acid corresponding to this anticodon on tRNA is attached to one of the free ends of the tRNA molecule, the molecule then being referred to as aminoacyl-tRNA, which is the "adaptor" molecule first postulated by Crick. Each adaptor molecule has to be specific in the sense of carrying the correct amino acid corresponding to the anticodon on the tRNA. A crucial enzymatic role in specificity is involved in insuring that this correspondence is correct when an amino acid is attached to a specific tRNA molecule. The adaptor molecules, specific for each amino acid, play an essential role in translating from the codon language on mRNA to the amino acid language for the protein.

The ribosomes in the cytoplasm provide the molecular machinery for carrying out the translation between languages referred to above. A highly schematic picture of this activity is shown in Fig. 8. In the diagram, the mRNA threads through the ribosome, one codon at a time, in the direction indicated by the small arrow (left to right). Molecules of tRNA are represented as rectangles and amino acids as small circles. The diagram indicates three successive tRNA's involved in reading successive codons on the mRNA message. On the right, the tRNA has been released from the active or recognition site after delivering its amino acid proline to the polypeptide chain (corresponding to codon CCU on the mRNA and anticodon GGA on the tRNA. See Universal Code Table I, Chapter 5). The central tRNA is now at the active (recognition) site with anticodon CCC pairing with mRNA codon GGG, with the glycine amino acid being removed and attached to the peptide chain. On the left, a third transfer RNA with anticodon UUG is ready to move into the recognition site to pair up with the next mRNA codon AAC with its appropriate asparagine amino acid attached. Enzymes also play an important role in this action at the ribosome whereby the appropriate amino acid is joined by a peptide bond to the growing protein chain after proper identification of the tRNA anticodon with the mRNA codon.

While all of this seems a bit fantastic, it is nature's clever way of dealing with the problem of ensuring that the correct proteins for the living organism are produced as specified in the genome.

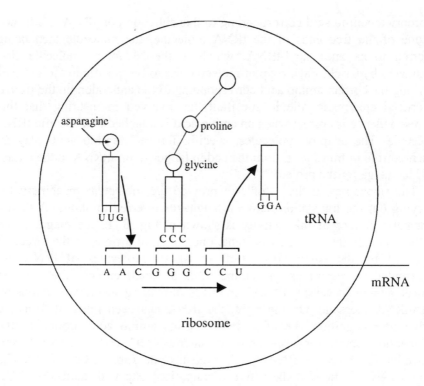

Figure 8. A schematic drawing showing the essential actions carried out by the ribosomes in the translation process. Imagine an RNA molecule with a specific succession of codons (three bases at a time) threading through the molecular apparatus of the ribosome in the direction indicated. The tRNA are represented as rectangles, the amino acids as small circles. When a particular codon on the mRNA occupies the recognition site in the ribosome, a search is made (effectively speaking) for a tRNA molecule with an anticodon which can pair (see Fig. 4) with the receptive codon in place. When a match is made, the tRNA molecule releases the attached amino acid so that it can be joined to the growing polypeptide chain (protein). Thus, the correct string of amino acids is put together in accordance with the instructions carried on the codons of the mRNA. See text for details. The correspondence between amino acids and codons on the mRNA is that in the universal code Table (Table I of Chapter 5). This is the translation process from a codon language to a language of amino acids.

4. Proteins—The Molecules of Life

Why are proteins sometimes referred to as the molecules of life? Certainly, they are not the only molecules involved, not even the only macromolecules involved.

DNA and RNA, just discussed, are obvious cases in point. But there are others as well, such as the lipid chain molecules that form cell membranes, both outer and inner, and various classes of polymers called saccharides made from starches and sugars. Then, of course, there is a great variety of small molecules, ions, amino acids and so on. But proteins have a special quality which allows them to perform many functions essential to cellular processes, and to such whole body processes as the immune system, the hormonal system, and the central nervous system. This special quality of proteins relates back to the details of their structure. Proteins are strings, both long and short, of amino acids bound together through peptide bonds. (Peptide bonds are formed when the amine group NHH on one amino acid attaches to the COOH carboxyl group on the next, with the release of a molecule of water. See Fig. 9 and Appendix A.) What makes them special is the fact that, out of all possible amino acids, only twenty are actually specified by the DNA in biological organisms (some additional amino acids are modified at incorporation, making the list eventually utilized greater then twenty). However, each of these twenty amino acids has distinctive properties determined by their residues discussed in Appendix A (e.g. they may be large or small; highly acidic or relatively basic; hydrophobic (avoiding contact with water) or hydrophilic (preferring contact with water)). These properties of the residues are related to their sizes and the electrical charge distributions on their molecular surfaces, leading to polarized molecules in some cases, neutrality in others. Depending upon the sequence of amino acids in the string, the properties of individual proteins are vastly different from one another. In aqueous solution, within appropriate pH ranges (38) and at physiologic temperatures, a protein string spontaneously folds up in a unique and characteristic way to give an overall form to the protein, which can be made use of in various ways in the system. In addition to the primary structure of a protein (just the sequence of amino acids themselves), proteins have secondary and tertiary structures having to do with how the string folds up, e.g., into a globular form, somewhat analogous to a tangle or ball of thick string. The string may fold in local regions in such a way as to form helical coils (the alpha helix coils (39) first discovered by Linus Pauling) or local sheets (called beta pleated sheets (40)), but all part of the same original string. In some cases, several independent strings associate with one another to produce a complex 3-dimensional form with a still higher level of structure.

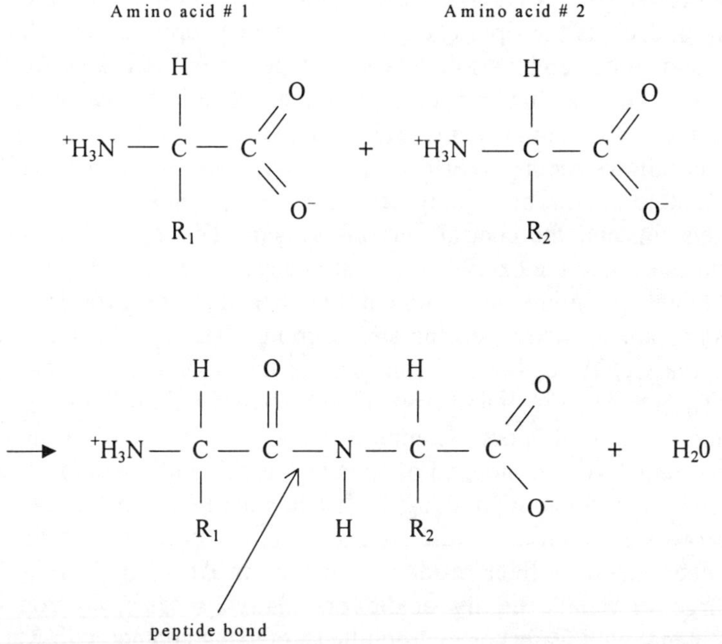

Figure 9. The formation of a peptide bond between two amino acids with the release of a molecule of water. A string of amino acids joined by peptide bonds is referred to as a polypeptide. Proteins are special polypeptides coded for in the DNA of an organism. Note that amino acids all have the same structure but are distinguished by their different residues, denoted by the symbols R_1, R_2, etc.

Since a protein (or its subunits) consists of one long string, there are special amino acid sequences which allow the string to go from a sheet region to a coil region and vice versa, retaining an overall globular form. But proteins can have a variety of other forms as well as the globular form, for example, some are long thin cylindrical shells, with an empty central core, made up from a special assembly of smaller proteins. This particular structure is that of a microtubule (41) which is a major component of cellular structure helping to give cells their form and rigidity.

The overall 3-dimensional form (conformation) of a protein achieved in the folding process is of crucial importance, as we shall see. Moreover, the conformation achieved may depend sensitively on ambient factors like temperature, pH, salinity and so forth. If a protein in solution is heated up considerably above physiological temperature, it will uncoil, losing its secondary and higher levels of structure (denaturation), but when cooled it coils up again, restoring its characteristic higher level conformation in a rather stable fashion. The essence of proteins is their great variety of forms (conformations), which can be exploited by the cell to carry out highly specific procedures.

Before pursuing this specificity further, let us first consider the question of how many proteins of a given chain length are possible, in principle. The answer is staggering, but gives some impression of the complexity of protein chemistry. Consider a polypeptide chain comprised of a string of just 100 amino acid units joined by peptide bonds (i.e. a moderately-sized protein), with units chosen at random from a pool of twenty amino acids. Since, in principle, at each peptide bond joining two successive amino acids, each amino acid could be any one of twenty choices, the number of distinctive strings of amino acids of this length is twenty raised to the power 100, an incredibly large number. In any particular cell in any arbitrary biological organism, only a small fraction of these, maybe none, may occur. It is up to the genetic code to specify which ones are required for that organism and to guide the protein assembling system to make these and only these, and, indeed, to make these only in the right amounts required—a daunting task, indeed. But this task is carried on in almost every cell of the many billions present in the human body every hour of every day, with almost no "fuss or muss" and without our even being aware of it, or for the most part worrying about it. Try to explain this by magic!! A lot of magic, to be sure.

But what makes proteins so important is not just that there is an incredible number of polypeptides to choose from to achieve any particular purpose. Rather, among all the possible polypeptides, some display an array of useful forms which make it possible to build just about any molecular machinery that one could conceive. And the greater wonder is that behind this impressive complexity are the simple laws of electrical forces which make it all possible. *Complexity from simplicity*, for the electrical laws are simple: just two kinds of charge (positive protons in the nuclei of atoms, negative electron clouds swirling about the atomic nuclei), like charges repelling and unlike charges attracting. Introducing some quantum mechanics, we account for all atomic structure and the so-called exchange forces (42) which operate between atoms and molecules. The basic law of electrical attraction is Coulomb's inverse square law (43), identified more than 150 years ago. Little did Coulomb realize then how his simple laws would play such a profound role in determining atomic and molecular structure, thus providing the physical basis for all biological structures and functions.

To understand the electronic structure of atoms and to see how atoms can bind together in various subtle ways, one needs quantum mechanics. But sufficient for our purpose is to know that atoms can join to form molecules, and some molecular units can string together to form long polymer-like strings, with complicated structures like proteins, and sophisticated structures like DNA and RNA.

When protein structures form, they often have electrical pockets or bumps on their surfaces, called receptor sites (11). If another molecule in the cytoplasm (referred to as a substrate) is shaped right, its surface may have a bump or a projection, or alternatively its own pocket, with a surface that fits the pocket or projection on the protein, and is consequently bound there by weak forces of attraction, the complex behaving in a kind of lock and key arrangement as illustrated in (Fig. 10). Fig. 10 goes a bit further to indicate the possible action of an enzyme in promoting the attachment of two substrates. In this instance, binding of the first substrate to the enzyme, modifies the first substrate so as to provide a binding site on it for the second substrate. The action of interest is the binding of the two substrates. Presence of enzymes can speed up the rates of reaction between substrates one and two by factors of many thousands. This is only one example of the action of enzymes. It is no exaggeration to say that life without enzymes is impossible to conceive, because many of the chemical reactions essential to the life of the organism proceed too slowly in the absence of enzymes or even do not proceed at all.

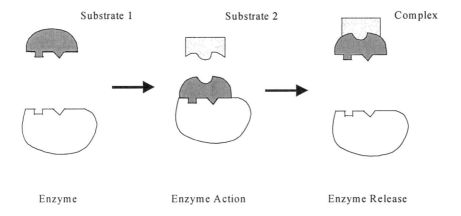

Figure 10. A schematic representation of one instance of "lock and key" binding between two substrate molecules assisted by enzyme action. In this instance the requirement is to bind two substrate molecules together. Left to their own devices, the process is very slow. The enzyme has a receptor site which binds effectively to substrate number one. This binding modifies the structure of the first substrate so as to create a binding site advantageous to substrate number two, so a complex consisting of the two substrates results. The enzyme is released in the process and is free to participate again in bringing substrates one and two together. Receptor site theory is basic to the understanding of biological processes in any organism. Enzymes, for example, speed up molecular processes by factors numbering in the tens of thousands or even in the millions in some cases.

All kinds of scenarios are possible. For example, a small molecule docking on one receptor site of a large macromolecule can so distort the electrical charges in the combined structure that another molecule which has already docked at another receptor site is released, with some appropriate consequences.

Receptor site theory is at the basis of almost all modern biochemistry, from enzymatic action which can speed up chemical reactions by factors of many thousands, to hormonal involvement in the endocrine system, to neurotransmitters which assist message transmission in the central nervous system, to highly specific antibody activity in the immune system. What all these processes have in common is great specificity of action and short reaction times.

The genetic program must be capable of passing on the genetic information from parents to offspring to enable them to grow and develop, with variability coming largely from sex through the sharing of parental genes in the offspring (44), nonetheless possessing the common features characteristic of their species. How this comes about is discussed in the next chapter, dealing with the discovery and nature of the genetic code.

5. Pinning Down the Code

Gene theory originated with the breeding experiments of the monk, Gregor Mendel (45), on the hereditary characteristics of ordinary garden peas [62]. His fundamental work was rediscovered in the early years of the twentieth century, but it was not until chemical analysis was applied to the problems of heredity in the nineteen forties, particularly by Beadle and Tatum [8] and subsequently by many others, that the existence of genes as the fundamental units of heredity was firmly established. Beadle proposed that genes are "nucleoproteins" which act as templates for the non-genic proteins, and that mutations correspond to chemical alterations in the DNA. This work was done using the fungus *Neurospora crassa* (associated with bread mold) which allowed metabolic control and identification of mutants leading to substantive results on the connection between enzymes and genes as fundamental units of heredity.

Another important step (and there were many) was the discovery by Spiegelman (46) [85], by Hershey and Chase [41], and others that, after a viral infection, the RNA synthesized in a bacterial cell was complementary to the viral DNA rather than to the host bacterial DNA. In other words, the virus had taken over the reproductive operation of the cell for its own purposes! It now seems clear that the virus construction allows it to inject its DNA and the necessary proteins (enzymes) for taking over control into the bacterium; one of these enzymes is lysozome, which disrupts the bacterial cell at the appropriate time allowing the newly generated viral progeny to burst forth from the cell, so that the process of infection of a bacterial colony is promoted and extended (see [86]).

Preceding the whole DNA story was the discovery of chromosomes and their apparent roles in determining hereditary characteristics. Chromosomes in cell nuclei could be seen under light microscopes, and individually identified by their overall structure and banding characteristics brought out with the use of suitable dyes. Clearly the chromosomes were involved in inheritance but the microscopic details were unknown. Attempts to make this identification dominated gene theory in the decades leading up to the discovery of the structure of DNA molecules. The rapidly growing interest in gene theory was put into "high orbit" by the discovery by Watson and Crick [95] of the chemical structure of the DNA molecule as a double helix consisting of two helical chains held together by molecular bridges (or "rungs," the whole having the appearance of a twisted helical ladder), as discussed in Chapter 3.

The now famous base-pairing rules of Watson and Crick (A with T and C with G) immediately suggested how, in principle, *replication* (32) of the DNA molecules was accomplished upon cell division, so that each daughter cell again carries in its nucleus the full complement of genetic information. But the mechanism for storing information in DNA and for utilizing it had yet to be elucidated.

It is of historical interest that the next step to Mendel in making an association between genes and inheritance was probably made by the British physician, Sir A.E. Garrod [32], when he studied the hereditary aspects of the disease Alkaptonuria. The concept of one-gene/one-enzyme was born. It was reasonably clear from the structure of DNA that the detailed information for genetic purposes could not be contained in the rather repetitious structure of the ladder sides themselves, but must reside in the sequence of bases forming the rungs of the ladder (the variable part) of the DNA structure. But the specifics of how the genetic information was stored and utilized remained a challenging mystery.

It gradually became clear from many studies in biochemistry that proteins were key to the specification of life itself, as discussed in Chapter 4. Although the work of Garrod was mostly unknown and unappreciated for some time, the Nobel laureate George Beadle referring to his historic work [8] dubbed him as the "father of chemical genetics." The implication seemed clear that the genetic information necessary for stable inheritance was the accurate specification of protein production in each cell separately—what proteins were necessary for any given cell in any given organism, in what quantities they were required, and in what time frames they had to be produced. In short, what was the protein library for the organism, and how was it employed? Further, how was the production of individual proteins regulated in the complex background of hundreds of chemical reactions in the cellular protoplasm? In the colourful phrase of Sydney Brenner, who is credited with the name "codon" for the triplet of bases coding for an amino acid in the genome [51, p. 469], this was the information required "to build a mouse," or any other organism for that matter [10].

The recipe for making any protein was eventually well established [14]. It was a matter of arranging in a linear chain the appropriate sequence of amino acids linked by peptide bonds, where the pool from which the amino acids are drawn is a special group of twenty amino acids occurring in biological organisms. So the problem of how, in principle, "to make a mouse" was to insure that the correct protein chains were constructed in the cells of the organism, with due regard for the fact that all cells in a multicellular system are "differentiated" (18) into cellular types, with their own, special protein requirements.

Back to the basic question: how was this information recorded in the structure of DNA, how was it *transcribed* (12) on to messenger RNA, and eventually *translated* (13) into protein structure in the ribosomes of the cell, transposing, in effect, from the nucleic acid language of base sequence on the mRNA to the protein language of amino acid sequence on the protein under production [68]? Whatever the mechanisms employed by the cell, they would have to include methods for recognizing where the information for a particular protein (gene) started and stopped on the DNA, and how the peptide bonds between the appropriate amino acids were formed in the ribosomes, all of this consistent with the fundamental processes of mitosis (47) and meiosis (48) in eukaryotes, involved with insuring stable inheritance on the one hand, yet allowing variability to arise within species on the other.

These remarks pertain to diploid systems, i.e. systems that have two sets of homolologous chromosomes. Heredity in haploid systems like bacteria (with only one set of chromosomes) follows a different set of procedures.

In one strategy of simplification, one could think of dividing the "mechanism problem" into two phases. (One has to bear in mind that the biophysical and biochemical processes involved in these mechanisms could not be directly observed since they take place at the molecular level. As a consequence, very clever experiments were necessary to track down the details.) Static pictures of molecular structures were sometimes possible with the aid of the electron microscope (49), but time-lapsed dynamic photography of any kind was unusually difficult and generally impossible, since significant molecular events making up biochemical reactions take place in time frames of the order of fractions of a millisecond. Much early analysis was carried out on phages (50).

In the two-phase attempt at simplification proposed above, one could think of dividing the mechanism problem as follows. In the first phase, one had to determine the nature of the coding process itself on the DNA. In the second, one had to discover the specifics of how the DNA information is translated into amino acid information on the protein under construction. In this first phase the primary work was done under Crick's direction at the Laboratory for Molecular Biology at the Medical Research Council in Cambridge, UK.

Crick [17] was able to determine the general nature of the coding strategy on DNA, namely, through triplets of successive bases (codons) forming the rungs of the DNA ladder. Crick also intuited the essential existence of "adaptor molecules," which were eventually determined to be a form of RNA called transfer RNA (tRNA) which could effect the translation process from triplet base codons to amino acids. Each adaptor molecule had to be very specific in the association of its anticodon on the tRNA and the amino acid attached to it.

In the second phase, the specifics of the translation process were gradually established, i.e. what specific DNA codon (base triplet) translates for what specific amino acid in a protein.

Although many experimental groups contributed in one way or another to the verification of the code, the most detailed work leading to specification of the genetic code table (see below) was done by Marshall Nirenberg's group at the National Heart Institute of the U.S. National Institute of Health in Bethesda, M.D. [66] and by Severo Ochoa's group at New York University [40, 67]. Parallel work confirming these results was carried out in Khorana's laboratory at the University of Wisconsin [54].

It is instructive to examine in some detail the experiments carried out by the groups under Crick and Nirenberg because they illustrate the kinds of "clever experiments" referred to above, which had to be devised to tie down the information which could not be obtained directly. It is, perhaps, appropriate here as an aside to make a comparison between the methods of theoretical physics and those of theoretical biology. In the former case, one is dealing with much simpler systems, in general, where the experimental results can be expressed through model-dependent mathematical equations whose rather precise predictions have then to be tested against whatever experimental observations are relevant. It is interesting that Crick, perhaps the leading scientific figure driving the discovery of the genetic code, had himself a background in theoretical physics which certainly helped in analyzing the basic challenges and in deciphering the genetic code; but he recognized that the methods of experimentation on the biological systems were much more complicated and indirect than the usual experimental methods carried out on physical systems. In the latter case, it was usually more obvious what the direct procedure would be to obtain the information required. This does not gainsay that sometimes the experiments in physics were also very involved, e.g. in high energy and particle physics, where often new equipment must be invented and employed. But in the biological situation, while much of the equipment was adapted from the physical and chemical laboratories, the methods of getting results tended to be very indirect, requiring great ingenuity and clever inference. The methods used by Crick and Nirenberg and their associates in phases one and two of the search for the code, illustrate the intense logic and ingenuity of the investigative process in biochemical experimentation leading to the nature of the code and its utilization—all this despite the general absence of elaborate mathematical equations, so much the normal arsenal of the theoretical physicist.

In phase one, the Crick approach was to study the class of mutants (51) leading to non-functional genes for the phage T4 (bacterial virus) infecting the intestinal bacterium *E. coli*. The essential experiments involved crossover (52) and recombination (53) between different mutant strains of the virus. The technique was to establish by crossover how functional genes could be constructed from two or more nonfunctional mutants in combination. The mutants corresponded to DNA with a base pair missing (designated a "minus") or a base pair added (a "plus").

This procedure allowed Crick to establish several important aspects of the nature of the code, in particular, that the codon was a triplet of successive bases, that the code was non-overlapping (54), that there was only a small amount of "nonsense coding" (i.e., few punctuation marks) and that there was a high degree of degeneracy (55) in the codon specification itself. The importance of selecting the correct reading frame was firmly established. This can be seen at once by examining an arbitrary sequence of bases in DNA. For example, suppose the string of bases is ACCTGCGATCCGA... In codon decomposition, starting with the left-hand side of the string, one would have the codon string (ACC) (TGC) (GAT) (CCG) (A...)... If one started in a reading frame one base to the right one would have (...A) (CCT) (GCG) (ATC) (CGA)..., while starting two bases to the right would give (...AC) (CTG) (CGA) (TCC) (GA...)... Clearly the string of codons in each case would be different, and one has to know which "reading frame" for codons is correct in order to code for the correct amino acid sequence for any given protein.

Assuming point mutations (one base deleted or added), and reading left to right, it is clear that a single "plus" mutation at some point on the DNA string would shift the reading frame, so that all subsequent codons to the right would be incorrect. Similarly, two adjacent "plus" mutations would alter the reading frame in between the two mutation points on the DNA, and again to the right of the second mutation. On the other hand, three "plus" mutations would restore the correct sequence to the right of the third mutation, so that normal coding function would occur past this point. On the other hand, a "plus" mutation followed by a "negative" mutation would restore the correct reading frame to the right of the "negative mutation." A similar analysis applies to "negative" mutations. The final test in these procedures is to assess whether function is restored or not in the crossover process. By this means, Crick was able to prove the conjecture (see Chapter 6) that the code was a triplet code, or at least some multiple of three; also that there was, in fact, a correct reading frame and that the code was not overlapping. Other general features of the coding process were also exposed by these procedures.

In phase two of the search for the code, one had to establish what codon corresponded to or coded for each amino acid. Nirenberg, Matthaei and associates [61, 65, 66] proceeded to establish the correspondence for each of the 64 distinct codons (three-letter words) that can be constructed from the alphabet of four letters, A,T, C and G. It turned out that three of the codons were "nonsense" signals corresponding to a STOP signal in the coding process. Since there are only twenty amino acids, clearly there is some degeneracy (55) in going from 61 sense codons to 20 amino acids. Ongoing investigations by various research groups established that the same code was essentially universal, i.e. that virtually all organisms use the same DNA code to specify amino acids, except for the degeneracy problem. (Rare exceptions will be discussed in Chapter 6.)

The technique used in Nirenberg's laboratory was to structure synthetic messenger RNA's from artificial polymers of nucleotides, in the first instance with poly U composed of successive units of uracil nucleotide in a cell-free system. In this system all 20 amino acids were present, but only one at a time was radioactively labelled with carbon-14. Clearly, in such a system, the synthetic mRNA carries only the codons UUU whatever the starting point for translation, so one would expect in a milieu of transfer

RNA's only those carrying the anticodon AAA would pair with the mRNA, and undergo transcription. One would then expect a protein product corresponding to a polymer of the amino acid corresponding to the UUU codon. The amino acid in the polymer was then identified by the radioactive labelling, a laborious procedure because separate experiments had to be carried out for each amino acid in turn radioactively labelled. This enabled one to identify which amino acid had been synthesized from the codon UUU. It was painstaking work, but highly specific. It turned out that the codon UUU coded for phenylalanine! Similarly, using poly A nucleotides for an artificial mRNA, it was shown that AAA coded for the amino acid lycine.

The experiments were then refined by making synthetic mRNA's from polymers with a repeating succession of bases, e.g., polymers of UA, UC, AC, AG, and CG. By cross checking the results, the whole code table (see Table I) could be deduced [51]. Gradually all the codons were sorted out, with the work of various laboratories contributing to the verification and universality tests. One interesting result implied that when the mRNA string passes through a ribosome, resulting in the production of a specific protein, it can then move on to another ribosomal site and repeat the process there, i.e. one mRNA string can move from ribosome to ribosome successively producing the same protein product on each.

In Table I, standard abbreviations are used for the various amino acids. The correspondence for each amino acid abbreviation is set out as follows:

Ala (alanine)	Asp (aspartic acid)	Arg (arginine)
Cys (cystene)	Asn (asparagine)	Gly (glycine)
Gln (glutamine)	His (histidine)	Glu (glutamic acid)
Ile (isoleucine)	Leu (leucine)	Phe (phenylalanine)
Met (methionine)	Pro (proline)	Lys (lysine)
Ser (serine)	Thr (threonine)	Tyr (tyrosine)
Trp (tryptophan)	Val (valine)	

The codon AUG is for methionine, which is involved in the signal required to start transcription of a segment of DNA to mRNA.

	U	C	A	G	
U	Phe	Ser	Tyr	Cys	U
	Phe	Ser	Tyr	Cys	C
	Leu	Ser	(Stop)	(Stop)	A
	Leu	Ser	(Stop)	Trp	G
C	Leu	Pro	His	Arg	U
	Leu	Pro	His	Arg	C
	Leu	Pro	Gln	Arg	A
	Leu	Pro	Gln	Arg	G
A	Ile	Thr	Asn	Ser	U
	Ile	Thr	Asn	Ser	C
	Ile	Thr	Lys	Arg	A
	Met	Thr	Lys	Arg	G
G	Val	Ala	Asp	Gly	U
	Val	Ala	Asp	Gly	C
	Val	Ala	Glu	Gly	A
	Val	Ala	Glu	Gly	G

Table I. The "Universal" Genetic Code. The base occupying the first position on the codon is given in the left hand column, that in the second position is given in the row across the top, and that in the third codon position is given in the right hand column (adapted from Alberts et al. (Molecular Biology of the Cell, page 108, Garland, New York (1983)). The three codons labelled (Stop) do not code for an amino acid, but are used in the transcription process to indicate the end of the sequence on the DNA corresponding to the protein being transcribed. Note that U rather than T is used corresponding to the convention that the code is defined for mRNA rather than for its template anticodon on the DNA.

6. Description of the Genetic Triplet Code

Even before the experimental analysis of Crick's group, there was speculation as to what the code might turn out to be. The guiding principle in this speculation was, how does one arrive biochemically at the specification of twenty distinct amino acids. If one assumes that successive base pairs on the rungs of the DNA ladder code in some way for each amino acid, i.e., that some type of codon exists in the *transcription* process, one could then speculate on the length of a codon. If, for example, each base on the linear DNA chain coded specifically for an amino acid (codon of length one) this scheme would only account for four amino acids, since there are just four bases in the DNA molecule. If on the other hand, pairs of successive bases were codons, then there would be 4 x 4 = 16 distinct base pairs (four choices of base at each of the two locations in the codon), which is still not adequate to specify twenty amino acids. On the other hand, the number of distinct triplets of 4 bases is 4 x 4 x 4 = 64, which gives an overspecification for twenty amino acids, leading to a degeneracy problem. There were also problems from a purely theoretical point of view, whether the code might be overlapping (54), and whether it was read off just one side of the DNA ladder or off of both sides. Of course, as we discussed in Chapter 5, Crick and his associates were able to show experimentally that the code is, in fact, a triplet code, that there is no overlapping of codons but that there is degeneracy (55). Moreover, experiment established that the process of transcription from DNA to mRNA is directional, proceeding from the 5' end of the DNA chain to the 3' end, (where the 5' and 3' labels correspond to a standard numbering of carbon atoms in the nucleotides; recall Fig. 3).

Searching for the code whereby the proteins of life are specified was pursued both by experiment and by informed speculation. The theoretical physicist George Gamow, a distinguished figure in the worlds of nuclear physics and cosmology, accepted the mathematical challenge of how to get in a natural way from 64 codons to 20 amino acids. His proposal was very clever, and while it did not correspond to experiment and eventually had to be rejected [18], it did contribute to thinking about the problem, clarifying the constraints imposed on any proposal for a universal code. Gamow's ingenious "diamond code" [31] was an overlapping code and assumed that the polypeptide chains were synthesized directly on the DNA. Experiment proved otherwise. In the meantime, the code table was arrived at on the basis of experiment, eliminating the need for clever speculation. The code as now established, is presented in Table I of Chapter 5. In Chapter 8, we will return to the problem of how 64 codons lead in a natural way to the specification of the limited pool of just twenty amino acids which are coded for in biology. The question will be addressed there in the context of theories pertaining to the origin and evolution of the code itself.

Table I shows how, from any selected codon, one can identify uniquely the corresponding amino acid. Because of the degeneracy problem, one cannot use the code in the reverse sense, i.e., one cannot begin with an amino acid and predict the codon from which it was translated. For example, the amino acid serine is coded for by no less than six different codons. Which of the six codons is used for each serine amino acid in a given polypeptide sequence can only be determined by going back to the DNA molecule itself or its transcript on mRNA. This is again an illustration and a consequence of the "Central Dogma of Molecular Biology" which states that the information flow is from DNA to RNA to protein. The central dogma does not always hold rigidly. An example occurs of reverse transcription in certain tumor-causing RNA viruses, where a DNA polymerase is RNA-directed. The new DNA formed then becomes the template on which mRNA forms and directs the translation into proteins [5].

A number of observations can be made from an examination of Table I. Perhaps the most pertinent is the extent of the degeneracy question discussed above. Another is the observation that base positions 1 and 2 almost specify the code by themselves, and are crucial to the specification, in general. For example, consider the amino acid serine, which has a six-fold degeneracy. In four cases, the first two bases in the codon are U and C while the third base can be any one of U, C, A or G. In the two other cases specifying serine, the first two bases are A and G, while the third can be either U or C. Another example is the amino acid leucine, which has again a six-fold degeneracy. In all six cases the second base in the codon is U. The first base is U in two cases, but the third base has A in one case and G in the other. For the other four cases the first base is C, the second U again, while the third can be any one of U, C, A, or G. In the six-fold degeneracy of the amino acid arginine, four of the cases have the first two bases in common, namely, C in the first and G in the second, while the third can be any one of U, C, A, or G. In the four-fold degeneracies of proline, threonine, alanine, valine and glycine, in each case the amino acid is already specified by the first two bases: CC for proline, AC for threonine, GC for alanine, GU for valine and GG for glycine. For each of these five amino acids, it is immaterial which base occupies the third place. Clearly the first base is, in general, more critical for specification than the second which is more critical than the third.

Another interesting point relates to point mutations. Most observed mutations of the virus T4 used by Crick in his studies were point mutations, i.e., mutations involving the addition, deletion or substitution of a single base. Substitutions in the third codon position are of no consequence at all for all of the amino acids with four-fold degeneracy, and for four of the six codons for all amino acids with six-fold degeneracy.

It appears on detailed analysis [92] that for most substitutional errors occurring in nature the amino acid replacement has much the same properties as the appropriate amino acid, so that the proteins produced are still functional, but with reduced functionality.

Finally a word about universality. Many organisms have now been studied and the results indicate that the genetic code is essentially universal. However there are some exceptions occurring in both yeast and mammalian genetic codes for mitochondria (56). Table II shows the differences which occur, as deviations from the universal code.

Codon	Mammalian Mitochondrial	Yeast Mitochondrial	Universal Code
UGA	Tryptophan	Tryptophan	STOP
AUA	Methonionine	Methonionine	Isoleucine
CUA	Leucine	Threonine	Leucine
AGA	STOP	Arginine	Arginine
AGG	STOP	Arginine	Arginine

Table II. Differences between the so-called universal code and two mitochondrial genetic codes. In each case the amino acid or STOP signal corresponds to the codon listed in the first column. (*Adapted from [2] Table 9-4, page 543.)

Mitochondria are small, incomplete organisms living symbiotically as organelles in the cells of eukaryotic (57) systems and are thought to derive from ancient bacterial species. They occupy up to one third of the cytoplasmic volume and have their own DNA but require the cell's genetic apparatus for replication. The value to the cell is that mitochondria are the "batteries" of the cell, with the property of being able to convert the energy from oxidative processes in the cell to the production of ATP (adenosine triphosphate (58)) molecules. The ATP molecules are the energy transport molecules in the cellular system, supplying for example the energy for muscular action by virtue of their high energy phosphate bond (59). Fig. 11 illustrates in a schematic way the general structure of ATP. The high energy phosphate bond is between the last two phosphate units and the energy released when ATP is hydrolyzed to adenosine diphosphate is about ten kilocalories per mole.

Different cell types have their own form of mitochondria adapted to their own needs and purposes. The ATP molecule through its high energy bond provides a convenient energy supply in cellular systems and is easily transported between cells. The citric acid (or Krebs cycle (60)) is associated with mitochondrial function. It is discussed in detail in Chapter 9 of the book by Alberts *et al* (loc. cit.). Discovered in 1937, it consists of a series of chemical reactions and is a major mechanism in most cells for the oxidation of carbon compounds.

The fact that mitochondrial and yeast codons differ in only a minor way from those set out in the universal table (Table I in Chapter 5) is consistent with the view that all life derives from a common ancestor but minor deviations occurred in certain niches in the course of evolution.

Description of the Genetic Triplet Code 43

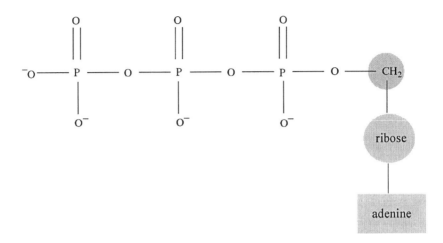

Figure 11. Schematic of the ATP (adenosine triphosphate) molecule. For steric reasons, the last of the three phosphate groups has a high energy bond with the rest of the molecule. Hydrolysis of ATP to form ADP (adenosine diphosphate) releases about seven kilocalories of energy per mole. For this reason ATP is an effective energy carrier in organisms, supplying energy to essential endothermic reactions.

7. Origin and Development of the Genetic Code

In previous chapters we have discussed the nature of the genetic code, how it is structured to elegantly store the information directing the processes essential to life and its inheritance. What we have not discussed at all is how this marvel of molecular machinery and ingenious process came about. However, the origin and development of the genetic code is but a part of the larger question. How did life itself originate? This larger question, which has been addressed variously over the ages, still has no conclusive answer.

Many conjectures and theories on the origin of life and development of the genetic code have been put forward, but none of them sufficiently definitive to wrap the matter up. A comprehensive, lucid and thought provoking survey of current conjectures and theories on the origin of life is given by Paul Davies in his book entitled *The Fifth Miracle* [20]. Davies' book not only explains what kind of ideas and theories have been put forward, but analyzes their various strengths and weaknesses. We confine ourselves here to a brief survey of the major problems that must be confronted by any theory on the origin of life.

First of all, even the experts disagree on the question, whether the origin of life is so improbable as to require a designer—a position put forward by M. Behe in his book *Darwin's Black Box* [9]. The counter-argument, relying on the pervasive acceptance of the Darwinian process of survival as effective designer, is put forward with great enthusiasm and elegance by Dawkins in *The Blind Watchmaker* [21]. The battle of words between proponents of these two points of view is taking place on the battleground of philosophy, since neither side has been able to establish their contention scientifically. Nonetheless, the weight of scientific opinion is clearly on the side that believes that life arose spontaneously out of inanimate matter over a long period of time from the distant past under relatively favorable environmental conditions.

Taking this predominant view that the origin of life can ultimately be explained by science, some fundamental questions must be addressed:

(1) What evidence can be adduced as to the physical conditions pertaining during the period when life first formed?
(2) In the words of Davies [20] is ours truly a biofriendly universe? That is to ask whether the laws of physics are or are not slanted toward the creation of living forms and to the evolution of increased complexity of living forms with time?
(3) Does life have a cosmic connection, so that it is created pervasively throughout the universe, or is its origin so improbable [45] that it might have happened only once, here on earth under particularly fortuitous conditions?
(4) How did nature solve the "chicken and egg" problem relating to whether nucleic acids preceded protein formation, or whether proteins came before nucleic acids? In the former case, how could genetic mechanisms operate without the presence of protein enzymes; in the latter case, how could proteins appear without the specification and regulation of a genetic mechanism, involving nucleic acids?
(5) How was self-organization (61) achieved? By the inevitability connected with autocatalyzing systems as suggested by Kauffman [53], or by the Prigogine and Stenger [70] processes of form and pattern development associated with open non-equilibrium systems driven by energy and material throughputs—or neither, or both?
(6) In any event, how did a genetic system of information storage and utility evolve from primitive life forms?
(7) Where does biological information come from? What kinds of information must have been available and in what forms to meet the requirements of any of the above processes? In other words, how can life form with the capacity to lower its entropy without some input of information?

It would appear essential to deal first with the question raised in point number one above, since only then can one deal reliably with the plausibility or otherwise of various chemical and physical processes hypothesized. Another starter stems from the universality of the genetic code. It suggests that there may be clues of primordial beginnings in the genomes of various organisms extant. One possible thread in this fascinating story is pursued in Chapter 8 of the present book. Consideration must also be given to the "chirality clue," the observation that biological molecules appear to be selected on the basis of chirality, e.g. the DNA molecule exists only in the form of a right-handed helix, never a left-handed helix, although the left-hand helices are not excluded by the laws of physics. (Handedness in helices is simply analogous to handedness in simple screws.) This observation suggests that all known life forms have evolved from the same origin. Otherwise, on a probability basis both forms would occur with equal probability.

In item five, the suggestions of Stuart Kauffman [53] and of Manfred Eigen and Peter Schuster [25] that life arose inevitably from systems of autocatalytic sets of chemical reactants provides a useful viewpoint, but no clue is given in such scenarios as to how a genetic system could have evolved. Moreover, no experiments in the laboratory have been carried out to confirm or deny the fundamental hypotheses involved in Kauffman's scenario. On the other hand, Prigogine's analysis shows that order spontaneously develops in non-equilibrium systems with a driving throughput of energy.

Perhaps the most provocative paradox is the so-called "chicken and egg problem": which developed first, a protein world or a nucleic acid world? It is provocative since in principle it should be possible to clearly resolve the paradox; and also because it relates directly to both the origin of life and to the subsequent development of the genetic code. The paradox concerns the twin questions, first, how could genetic systems evolve and operate without the extensive availability of proteins as essential enzymes and structural elements, and second, how could the highly specific enzymes necessary for the genetic processes be fashioned without a genetic specification and regulating system already in place?

Various theories and conjectures have tended to fall into the categories of proteins first, or nucleic acids first. Most popular in the latter category have been various versions of what has come to be called "the RNA world." Basic to theories of this type has been the observation that RNA has some autocatalytic properties, providing a possible basis for a reproductive scenario for primitive life forms without the aid of a battery of helpful enzymes. The RNA scenario, however, focuses on replication to the exclusion of the metabolism necessary for the propagation and evolution of living organisms (see [20]).

The alternative is to examine the possibility that proteins came first, a scenario explored particularly by the theoretical physicist Freeman Dyson

[24]. Dyson's ideas, with some correspondence to those of Kauffman, suppose that under the protection of Oparin-like cells, a primordial soup with a complex network of chemical interactions undergoes a spontaneous transition from a chaotic phase to an ordered phase, in rough analogy to physical phase transitions, such as water (chaotic) to ice (ordered).

Wald [93] gives some critical observations on the conjecture of improbability that life spontaneously formed from inanimate matter. Particularly significant in his remarks are examples to show how seemingly improbable individual events, with the aid of long time periods, can become probable, even certain. The strengths of considerations such as those of Kauffman [53], Dyson [24] and Wald [93] is their plausibility in the overall scheme of things, suggesting that they may play important roles in a final synthesis on how life came about. Skepticism about Wald's optimism on probabilities nonetheless persists. It is, perhaps, most colorfully phrased in the analogy of Fred Hoyle [45] that the generation of life as a series of steps, each of a highly improbable nature, has about the same odds as for a whirlwind sweeping through a junk pile to produce a fully functioning Boeing 707. This is part of Hoyle's argument that life probably came to earth from outer space.

The present state of current theories on the origin of the code is explored by Alberti [1] in a review article in the journal *Cellular and Molecular Life Sciences*. See also reference [7].

Alberti discusses the two traditionally opposing views on the evolution of the genetic code, namely, theories which are stereochemically based, and theories which can be characterized as "frozen accident" theories. The stereochemical theories predict significant affinity between codon and amino acid pairs [102]. This could provide a strong chemically based "driving force" or influence for the origin and evolution of the genetic code. However, there seems to be little direct evidence for this selective binding ([87, Chapter 27]). In a stochastic approach the frozen accident theory relies on random assignments of amino acids to codons. According to such theories, origins were a matter of chance rather than of necessity.

As to the "chicken and egg" problem, Alberti conjectures that nucleic acids and protein structures evolved together. He speculates that sequence-specific interactions between polypeptides and polynucleotides likely formed the core of a primordial genetic code. Crucial to his theory is the assumption that sufficient selective binding occurred between polypeptide and polynucleotide chains over significantly long regions of these molecules to provide stability of process in the high temperature environment characteristic of primordial cells or enclosures then available for the generation of life. One would then expect that the "core codons" associated with this circumstance would be much longer than the triplet base codons observed today. Evolutionary processes would then gradually reduce the codon lengths to the minimum (triplet base) codons of our present era. Alberti's theory assumes that the translation and transcription machinery would be essentially retained throughout this transition. The advantages of this theory are first of all avoidance of the chicken and egg paradox, and second, the possibility that its conjectures can be tested in the laboratory.

Davies [20] discusses the possibility that life began, not only under the sea, but even under the seabeds, which over time spread to the oceans and eventually to land organisms. According to this scenario, the first living organisms were archaea (62), single-celled pre-bacterial forms possibly related to the superbugs (63) observed today. Despite the wealth of imaginative concepts relating to theories on the origin of life, there are still more questions than answers, and a definitive synthesis seems a long way off. Perhaps the most persistent feature in the exploration of life and how the code came about is the universality of the code itself, suggesting that either all life has a single ultimate ancestor, or that influences leading to the very existence of life were pandemic, since the same code evolved wherever, whatever and whenever the generating circumstances were propitious.

8. A Physical Approach to Genetic Origins

In Chapter 7 we reviewed some current theories on the origin and development of the genetic code, and how these aspects of the code relate generally to the origin of life itself. It is evident that unravelling the mysteries of the origin of life and of the code are closely interconnected, and this process is bound to be a piecewise one, depending on ever more definitive biochemical experiments with guidance from the structure of the universal genetic code now extant, and from inferred information pertaining to the nature of the prebiotic and early-biotic environments. Biochemical experiments coupled with thermodynamic inferences help to delineate the possible chemical reaction scenarios that lead from prebiotic conditions to the origin of the first primitive biological organisms.

Various theories and speculations have been put forward as to the processes and conditions that may have significantly contributed to an understanding of the origins of life and development of the genetic code. In this chapter, we shall focus on one possible scenario with which the author has been involved (This work was surveyed in the book, *Physical Theory in Biology* [60, p. 405]. It is discussed here in a somewhat simpler form.) It is emphasized not because it has been proven to be correct in all its aspects, but because it raises some rather specific issues, which in time may well contribute to a more complete and final solution to this interesting problem. The analysis, which is presented in three stages, leads to two specific results which are interesting in themselves, whatever their significance for a final theory. One of these is the deduction from observed codon biases and from thermodynamic considerations, that organisms can be classified into particular groups which correspond to standard classifications (e.g. prokaryotes as opposed to eukaryotes (57), bacteria as opposed to viruses, etc.). The second is a very neat representation of "codon space" viz. what we shall call the *tetrahedral representation*. This representation is highly suggestive of some features of the origin of the genetic code and in particular clarifies the question of how the degeneracy problem (i.e. the problem that several distinct codons code for the same amino acid) came about, and why it is natural for 64 codons to represent 20 amino acids. Chapter 8 will be presented in three sections, which tell a story of discovery, each section representing a distinct step in the analysis.

The almost complete universality of the genetic code is in itself highly suggestive. It would seem on the one hand, that all organisms, plants, animals, yeasts, whatever, came from a common ancestor in a fortuitous set of circumstances; alternatively, life may have developed from more than one ancestor, but in each case, it would seem that the same set of favorable circumstances applied, that is to say the origins of life followed the same favorable set of biochemical pathways. In either case, the fortuitous circumstances included a web of complex biochemical reactions which, in turn, reflected favorable thermodynamic circumstances. The viewpoint taken in presenting the following scenario relies heavily on the probable importance of thermodynamics bearing on the origin of life and the closely associated problem of the origin of the code.

In the first section of this chapter, the degeneracy problem is considered. This analysis leads in Section 2 to a definition of "codon space" and to the remarkable clustering of points which occurs in that space. Finally, in Section 3 we are lead by these circumstances to a novel representation of codon space, which is in itself suggestive as to how the genetic code developed.

The scenario presented in the next three sections was originally published in a series of papers as follows: Rowe and Trainor [75]; Rowe and Trainor [76]; Rowe [74]; Rowe, Szabo and Trainor [77]; Trainor, Rowe and Szabo [89]; and Rowe [78].

8.1 Codon Bias in Viral Genes — A Thermodynamic Theory

In [74], an analysis of the informational content of DNA in several viral genomes was carried out using an extension of the analysis developed by Gatlin [33]. Viral genomes are relatively short, on the order of 10,000 bases. Sometimes they occur in circular rather than in linear form, and can consist in individual cases of either DNA or RNA. The number of genes packed into a viral genome is of the order of ten and overlapping of genes occurs in some circular forms. (Overlapping is where a string of bases contributes to more than one gene product.) The analysis showed that for coding regions (64), there is information structure in the genome at the level of single bases, pairs of bases, and particularly at the level of three bases; but there is no significant long range structure beyond three bases. By information structure we mean that information is contained in relationships or regularities between bases separated by any given "distance" along the genome. The result that there is information structure at the level of three bases is not, in itself, surprising from a physical point of view and had been noted previously by several observers. After all one would expect nearest neighbor interactions between these large molecules (nucleotides) to dominate which suggests information structure at levels two and three. How this may have played a part in the origin and development of the triplet code is suggestive but not yet clear.

A possibly significant observation coming from the analysis of [75] is that, by contrast, non-coding regions (64) (see Chapter 6) show a nonrandom base sequence often containing a higher-level structure than is found in the genes themselves.

In [76] this work was followed up by a consideration of the origin of the codon bias observed in viral genes. Since the code is highly degenerate (several distinct codons coding for the same amino acid as discussed in Chapter 6), it is natural to ask what is the frequency distribution of use in various genomes of the various codons for the same amino acid. For example, we saw earlier that the amino acid serine is coded for by no less than six distinct codons. In particular, for any given protein, when serine occurs, what is the determinant that selects a particular codon among the six choices to insure that serine is selected? The distribution of possibilities might turn out to be random or "flat." If not random, the selection of a codon will show a bias since some codons are then used more frequently than others.

The possible origins of such biases is discussed extensively in Rowe's thesis [74], and in references [76], and [90]. For one thing, a codon bias could lead to restrictions on the amino acid compositions in a protein specified by a particular gene. For example, codons of the form Gxx (where G stands for the base guanine in the first codon slot, and "xx" means any choice of the four bases in the second and third slots, like wild cards, as it were) code only for the amino acids Val, Ala, Asp, Glu and Gly (see Table I, Chapter 5). Val and Ala are hydrophobic, Asp and Glu are hydrophilic, and Gly is electrically neutral. From Fig. 5 in Dickerson [23] it appears that the amino acids specified by two of the codons of the Gxx form are likely to be found on the exterior surfaces of proteins (Asp and Glu), one is to be found predominately in the protein interior, and two are relatively ambivalent as to where they might sit. Thus from an observational point of view, this implies that a bias toward G in the first codon position is not very restrictive in the gross structure of proteins.

The bias towards codons of the form xAx and xCx (with preference for either A or C in the second codon position) is also not very restrictive. All xCx codons lead to weakly hydrophilic amino acids, and xAx codons code for amino acids of all types. The preference for U in the third codon position (i.e. xxU bias) is also not a strong constraint since only the amino acids Trp, Met, Gln and the Stop signal are excluded by this form. Aside from electric charge specificity, there may be other constraints on building a protein, such as amino acid sizes, but in any case this would not account for the xxU biases, particularly since the third base in many cases plays no part in amino acid determination. This is evident from an examination of Table I, Chapter 5.

A crucial constraint in the translation process is how the transfer RNA active site recognizes a specific codon site on the mRNA, since the particular tRNA binding to a codon site must carry the correct amino acid for any particular step in the protein assembly. The original speculation, supported by subsequent evidence, showed that accurate identification between the codon on mRNA and the anticodon on tRNA was achieved through Watson-Crick base pairing. This speculation required that a particular anti-codon on tRNA would be highly specific in recognizing one and only one codon on mRNA. While this was generally true, it was found not to be always true. For example, the alanine tRNA in yeast (first studied by Holley [44]) binds to any one of the three codons GGU, GCC or GCA. Crick [15] explained this in his *"wobble hypothesis"* as due to the fact that the steric binding between codon and anticodon pairing in the third codon position was less stringent for the third codon position than for the first and second codon positions. Crick's hypothesis was borne out by experimental studies.

But this analysis leaves unanswered the question why there is a bias toward U in the third codon position. Various attempts at an explanation have appeared in the literature, particularly by Grosjean *et al.* [38], who suggested that codon bias, at least in the phage MS2, may be connected with the stability of the codon/anticodon interaction, keeping in mind that a codon which does not have to use a "wobble hypothesis" mechanism to achieve an anticodon binding is favored since the wobble match is weaker than a true Watson-Crick match.

A Physical Approach to Genetic Origins 55

In [76] and [78] a different approach to the codon bias problem is taken, viz. rather than assuming that codon bias developed during the period of evolution of the code, codon bias may have preceded this evolution. This hypothesis intimates that codon bias was a strong determinant of how DNA was to be used biologically.

The basis of this hypothesis is the consideration that if DNA formed in the primitive environment, without enzymes present to direct its construction, even perhaps in a cell-free environment, it would polymerize in a manner which is very sensitive to the relative stability of each step taken. This relative stability would lead (molecular Darwinism (65)) to a favoring of certain forms. This hypothesis leads one to suggest that the structures of early segments of DNA were probably dominated by periodic polymeric forms (67), which were later used to organize DNA into the specific forms observed today.

To test this hypothesis in a substantive way, a thermodynamic analysis was undertaken to determine the most probable structures of early DNA segments using a modified Ising model (66), well known in physical theories of magnetism, and elsewhere in condensed matter physics.

To this purpose, consider a double-stranded DNA polymer forming in some appropriate prebiotic environment, without the aid of enzymes and other molecular conveniences, with only thermodynamics to guide it. This problem was modelled in [76] by defining an "element" at any particular "site" on the growing double-stranded DNA polymer as a nucleotide pair, i.e. as a Watson and Crick base pair with the appropriate sugar molecules attached. At any point in the polymerization process, the question arises as to what "element" is most likely to be added to the last site on the growing polymer? The "elements" as defined above already include Watson-Crick pairing (through the usual hydrogen bonds), so the choice of next element should be determined by the so-called stacking energies discussed by Volkenstein [92], which could be defined as the negative of the energy required to remove each element from the chain in the reverse process. The various "elements" competing differ only in whether their Watson-Crick pair corresponds to AT, TA, CG or GC base-pair bridging, and to which immediately previous "element" terminates the chain up to that point. Since there are four termination elements possible and four newly added elements possible, there are $4 \times 4 = 16$ different stacking energies to calculate. However, from the complementarity of the chain, it turns out that there are only ten rather than sixteen independent stacking energies [76].

Crucial to the model is to know what these ten stacking energies are, as these are the parameters which thermodynamically determine the relative probability of occurrence of the various "elements" in the DNA polymerization process. Unfortunately, stacking energies are not well known and recourse has been made in the literature to quantum chemistry calculations. Rein [71] published a table of best values, but these were not sufficiently definitive to be of much use. In the calculations of [76], the values quoted by Volkenstein [92] were used, but only as a guide in the following sense. The stacking energies and the temperature were treated as parameters in the statistical calculations, and a search made for each periodic primordial DNA sequence. The results were encouraging because the energy values so obtained corresponded well with the values quoted by Volkenstein.

The model used in [76] was essentially a four state Ising-like (66) model, assuming that double stranded DNA formed a linear chain. These various hypotheses are defended in [76] and in more detail in Rowe's PhD thesis [74]. Of course, the primary assumption is that primitive DNA segments were formed under conditions where statistical mechanical analysis could apply.

We shall not pursue the details of these calculations here, but only summarize the results. The basic outcome was to indicate that pre-biotic and early biotic DNA polymers were likely to form as *periodic polymers* (67), and that of these some were more likely to form than others. This approach would contend that modern DNA should have features reflecting this primitive origin. While experimental parameters are not sufficiently well known to make a definitive case, it does show that thermodynamic considerations probably played an important role in the development of DNA structure, and by consequence, in the structure of the universal code.

8.2   Cluster Analysis in Codon Space

Over the last few decades, rapid progress has been made in sequencing sections of DNA from a variety of organisms. In many cases entire genomes have been sequenced, as evidenced by the large DNA libraries, such as the Genbank at Los Alamos and the Dayhoff in Maryland[4]. These data relate not

---

[4] This kind of information is now available from the National Center for Biotechnology Information, National Library of Medicine, National Institutes of Health, 8600 Rockville Pike, Bethesda, MD 20894, USA.

only to the genomes of various species, but also to the sequences pertaining to individual proteins, ribosomal RNA's, transfer RNA's, and also to noncoding genomic sections of DNA.

It is clear that codon degeneracy is treated differently by different organisms (i.e. the codon biases are different for different organisms). These degeneracy patterns form an important probe as to species origin and placement on the tree of life. Grantham [36] has attempted to relate the various codon biases to the source or function of the gene with interesting results.

A clearer picture of these sequence correlations comes from the analysis of Rowe, Szabo and Trainor ([76, 77]). In this work, the authors defined a convenient 9-dimensional "codon space" which is simple to construct from the experimental data, and which reveals a surprising picture of how the genes of organisms cluster in this space.

The first step is to define a "codon space" in such a way that every DNA sequence, whatever its origin, can be represented as a point in this space. One can then use cluster analysis to see if any useful clustering occurs, and whether the clustering is species or organism specific.

Consider again the triplet genetic code. Each codon is a triple of three successive bases in the appropriate reading frame. We refer to the three positions for a base in each codon in the reading frame as slots 1, 2 and 3. A particular codon is then defined by which of the four bases (T, A, C and G) occupies each slot in the reading frame. For any given gene sequence, one can easily count twelve numbers, viz. the number of each of the four bases occurring in each of the three slots. Let these numbers be designated with two characters, the slot number i running from 1 to 3, and the base number x running from 1 to 4, according to which of the four base populations are under consideration, i.e. x takes on the values T, A, C, or G. (Note that with the convention that codons are defined with respect to mRNA rather than the DNA template, the codons referred to here are strictly speaking anticodons, but we shall refer to them as "codons" for convenience.)

Thus for any sequence of DNA, we can construct a 3 x 4 matrix of numbers as follows:

$$M = \frac{1}{N}\begin{pmatrix} T(1) & A(1) & G(1) & C(1) \\ T(2) & A(2) & G(2) & C(2) \\ T(3) & A(3) & G(3) & C(3) \end{pmatrix} \quad (8.1)$$

where $A(n)$ is the number of times that the base A occupies the nth slot in the codons in the DNA sequence (n = 1, 2, 3), and similarly with the other base populations. The factor $1/N$ converts these populations to fractions of all N nucleotides in the DNA coding sequence. In equation (8.1), only nine of the twelve populations are independent variables, because the sum of elements along any row is fixed at $N/3$, so we can reduce the matrix to a 3 x 3 matrix of 9 independent elements by dropping one row, e.g. the $C(n)$'s in matrix (8.1). Equation (8.1) can then be replaced by

$$M = \begin{pmatrix} t(1) & a(1) & g(1) \\ t(2) & a(2) & g(2) \\ t(3) & a(3) & g(3) \end{pmatrix} \qquad (8.2)$$

where $a(n) = 3A(n)/N$ is the fraction of nth slot occupied by base A. Similar definitions apply to the other elements of the matrix (8.2). Division by $N/3$ tends to put long and short sequences on the same analytical basis.

The nine independent matrix elements in Equation (8.2) allow us to plot the whole DNA sequence as a *point* in nine-dimensional *codon space*. Thus one can take any known DNA sequence and represent it as a point in codon space. A refinement of this procedure, which was carried out in [76] is discussed in more detail in Appendix B. The simple approach described in Appendix B revealed a remarkable degree of clustering. It is difficult to assess what the clustering signifies in a nine-dimensional codon space, so three-dimensional projections were made from the nine-dimensional space, for example onto the A(2), T(2), G(2) subspace as shown in Fig. 12. Naturally, information is lost in such projections, which are made primarily for ease in visualization. For example, two points belonging to different clusters in nine-dimensions, may not appear to be so clustered in a two dimensional projection. It turned out that for projections onto the three-dimensional subspace with axes A(2), T(2) and G(2) illustrated in Fig. 12, little information is preserved in the projection, whereas for projections onto the first and third codon positions, meaningful partitions of space occurred, even in three dimensions.

In general, the distribution of points corresponding to protein coding sequences is highly non-random. Cluster analysis reveals a tendency for "the genes of more primitive organisms" to lie further from the origin in A, T, G-space than the nuclear genes of higher organisms. Cluster analysis clearly

A Physical Approach to Genetic Origins

shows that clustering in codon space suggests a useful approach to understanding subtle changes in the evolution of genetic sequences. But more on this later.

The work of Rowe and Trainor [76] demonstrated the importance of the concept of "codon space" and of the clustering of points therein (i.e. of the distributions of the points representing DNA sequences in this nine-

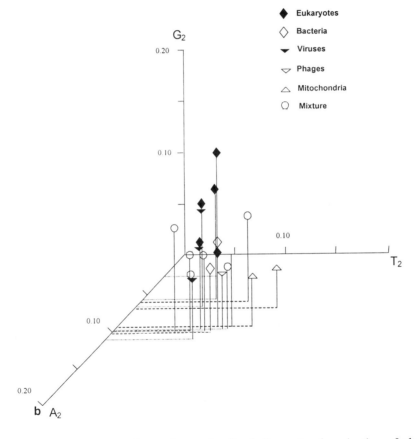

Figure 12. This figure shows the result of a 3-dimensional projection of cluster points from a 9-dimensional codon space where the axes of the three dimensions have been chosen as the numbers $A(2)$, $T(2)$ and $G(2)$, i.e. the A, T and G populations in the second slots of the coding frame (see text). Points correspond to the tips of the vertical lines, the latter being used for coordinate locations only.

dimensional space). Rowe [78] then pursued the question of what this clustering might signify. His approach was to use the technique of multi-dimensional scaling (MDS) developed by Kruskal [56, 57], which had been used by French and Robson [29] in another context, namely in investigating the biologically significant properties of amino acids in a genetic context.

The MDS of Kruskal tackles the following problem in a systematic way. Suppose one has a distribution of points (locations) in some higher dimensional Cartesian space, with a "distance" defined between each pair of points. MDS then addresses the question whether these points could be represented in a space of lower dimension in a more revealing way. (Examples come to mind: In two space, a distribution of points might fall on a straight line, in which case a one-dimensional representation is appropriate. It also could turn out that a distribution of points in three-space might fall in some plane, in which case an appropriate two-dimensional representation is most revealing.)

Rowe used this scaling procedure to see whether the information contained in the nine-dimensional distribution in codon space, might usefully be represented in a space of fewer dimensions. Kruskal [56, 57] had previously suggested that one should have some measure as to how "stressful" this compression was turning out to be in specific cases. He established a mathematical definition of "stress" for each application and adopted a criterion that the stress level should be less than 0.11 on a scale of 0 (no stress) to 1. (The stress tells how difficult it is to make the compression.) As shown in Fig. 13, Rowe, using Kruskal's procedures, calculated the stress involved when the nine-dimensional codon space distribution was compressed to 8, 7, 6, 5, 4, 3, and 2 dimensions in succession. Fig. 13 shows clearly that the optimum is three, as this is the lowest dimension which satisfies Kruskal's criterion that the stress should be less than 0.11 for a meaningful projection, and one is trying to achieve the lowest possible dimension.

To summarize: Given any DNA sequence, it can be represented by a point in codon space. When one does this for a large number of sequences from different organisms, one gets a clustering in codon space, which turns out to have significance in terms of parameters in a specific three-dimensional space. What these parameters are is not part of the procedure and must still be determined from other considerations.

Clearly, something is going on here of evolutionary significance since the clustering correlates with such factors as the evolutionary ages of species, whether species are prokaryotic or eukaryotic, whether they are mammalian or non-mammalian, or whether they belong to special classifications designated as mitochondria, viruses, bacteria or phages. What Rowe has shown is that the significant differences between clusters already show up in some appropriate three-dimensional compression of nine-dimensional codon space. One anticipates that a further compression from three to two dimensions can still show interesting trends. (See, for example, the two-dimensional projections shown in Fig. 12 above.) We note, however, in the optimum projection of Kruskal, only three parameters then suffice to classify organisms into one of the five major groups (mitochondria, viruses, phages, bacteria and eukaryotes). These parameters derive from the base compositions of the various gene sequences and are roughly expressible in terms of them. The base compositions, however, are likely only one manifestation of the thermal, chemical and biological influences which position the sequences in their relational locations in the three-dimensional MDS space indicated here.

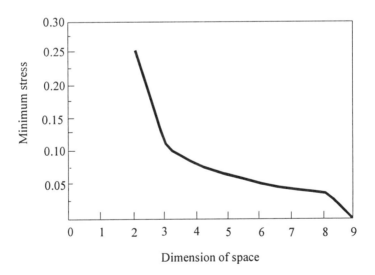

Figure 13. Reproduction of a graph by Rowe [78] with permission. The graph plots stress against reduced dimension when one compresses data from a higher to a lower dimension. The stress is a measure of how severe this projection is.

8.3 The Tetrahedral Representation of Codon Space

Section 8.3 is primarily concerned with a representation of codon space [89] which is geometrically attractive, which appears to take the mystery out of the question of getting from 64 possible codons in a triplet code to the specification of the 20 amino acids occurring in biology, and finally, which bears significantly on the question of the origins of life and of the code itself. This novel representation does not put forth new information so much as it displays previous information coming from the analyses of Sections 8.1 and 8.2, which make associations between codons and amino acid determinations more transparent and suggestive. The tetrahedral representation is described in detail in the following paragraphs, but suggestions for actually constructing a three-dimensional model are left to an Appendix (Appendix C – *Fun with Tetrahedra*).

The codon vocabulary is 64 three-letter words that can be formed using a four-letter alphabet, namely A, T, C and G for DNA, alternatively A, U, C and G for RNA. We have discovered [89] a representation of the 64 codons which is geometrically attractive and stimulating.[5] This representation was probably inspired by an analogous kind of representation used for elementary particle quantum numbers in modern particle physics (see, for example, the article by Frank Close *The Quark Structure of Matter* [13] in the book *The New Physics* edited by Paul Davies [19]). Our tetrahedral representation can be described as follows:

> A regular tetrahedron is a solid with four identical faces, each face being an equilateral triangle, as shown in Fig. 14. Label the four vertices with the codon triplets AAA, UUU, CCC and GGG, as indicated in Fig. 14. Since all vertices are geometrically equivalent, the order of labelling is arbitrary at this point. Choose any one of the faces of the tetrahedron as the geometrical base of a three-sided pyramid. For example, the choice of pyramidal base could be the triangle, UUU, GGG, CCC shown in Fig. 14. Imagine now passing three planes parallel to this base, with the top plane passing through the apex AAA of the pyramid, and the other two equally spaced between the base and the apex. (The base itself defines a fourth

---

[5] The reader is reminded of what was noted previously, that in Section 8.2 we referred to codon space with respect to DNA. Here we revert to the conventional reference to RNA, hence to the letters A, U, C and G rather than A, T, C and G.

A Physical Approach to Genetic Origins

parallel plane.) The second and third of these planes, of course, intersect the tetrahedral faces along a number of lines.

From the symmetry of the tetrahedron, we see that this geometrical construction has four variants. The variant chosen above has AAA as an apex, and the triangle UUU/GGG/CCC as a base for the pyramid, but we could just as well have chosen UUU as an apex, or GGG as an apex, or finally CCC as an apex. In each case, the construction of a set of four parallel planes could be carried out in the same manner and one would have the intersections of these planes with the other tetrahedral faces.

As we can see from Fig. 15, specialized to the face defined by vertices AAA, UUU and CCC, each face of the tetrahedron is divided by this construction into nine equilateral triangles, six of which have apexes pointing in the same direction, say "up," and three pointing in the opposite direction (down).

The net result of the above construction is to create a tetrahedral lattice with twenty points on the surface of the tetrahedron, with no internal points, defined by the intersections of the twelve planes defined above with the

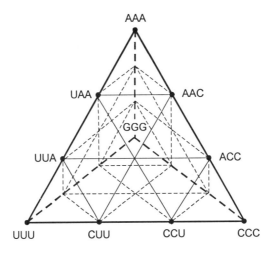

Figure 14. This figure depicts a tetrahedron with the four apices labelled U, G, C and A. As well it shows the intersections of the various planes parallel to successive faces of the tetrahedron with other faces of the tetrahedron (as described in the text). These intersections will be identified with codons and eventually with "codon families"

surfaces of the tetrahedron. A suggestive labelling of the points (to be replaced shortly by a more perspicuous choice) can be illustrated by reference to the face AAA,UUU, CCC shown in Fig. 15. Starting from the vertex labelled UUU, and moving to the right in the direction of vertex CCC, it might seem natural to label the next intersection by dropping one U in the UUU codon and replacing it with a C so that this intersection is labelled with the codon UUC, for example. But an equally valid candidate could have been UCU or CUU. (This is one reason why we consider a relabelling of intersections below.) In any case, moving to the second intersection between UUU and CCC, one would be tempted to label it with a codon in which two U's have been replaced by C's, for example UCC, or CCU or CUC. Similarly, moving from UUU toward AAA, we might expect to gradually

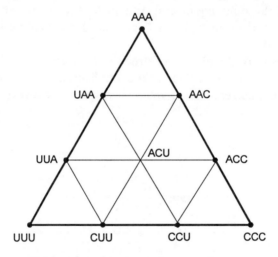

Figure 15. To simplify the results of the construction leading to the representation in Fig. 14, we consider taking one face of the tetrahedron at a time. Fig. 15 displays the AUC face which shows the intersection points of the planes described in the text with this AUC face. Counting all four faces there are just twenty intersection points altogether, with no interior points, which can be labelled with the twenty families of codons set out in Table III of Chapter 8. In this way, the 64 possible codons in the universal genetic code table can be grouped into twenty families, each family corresponding to a codon and its non-equivalent permutations, and each represented by an intersection point on the faces of the tetrahedron.

A Physical Approach to Genetic Origins

drop U's in favor of A's, whereas moving from AAA toward CCC, we would drop an A and pick up a C at each successive intersection point.

Finally, at the center of the face shown in Fig. 15, there is a unique point for that face, here labelled ACU, corresponding to the intersection of three sets of planes with the ACU surface of the tetrahedron. The labelling is suggestive because one reaches it from vertex AAA, say, by moving two steps in the C direction, followed by one step in the U direction. For this special codon designation, however, there are 6 codon choices, ACU, AUC, UCA, UAC, CAU, CUA all belonging to the same codon family (see Table III).

Letter set	Codons in the family
A, A, A	AAA
C, C, C	CCC
U, U, U	UUU
A, A, U	AAU, AUA, UAA
A, A, C	AAC, ACA, CAA
1C, C, A	CCA, CAC, ACC
C, C, U	CCU, CUC, UCC
U, U, A	UUA, UAU, AUU
U, U, C	UUC, UCU, CUU
A, C, U	ACU, AUC, CAU, CUA, UAC, UCA

Table III. Codon Families associated with vertices on the tetrahedral lattice.

We are now ready to consider a refined relabelling, consistent with our construction, but with content relating to previous considerations as well as to some experimental observations: rather than choosing say UUA over UAU or AUU, as possible codons with two uracil bases and one adenine base, we let this intersection point represent all three. In other words, each intersection point on the faces of the tetrahedron represents a codon and the codons formally obtained from it by permutation of its base composition. In Table III, we set out the codons arising from any particular choice of the three bases in the codon. We refer to these equivalents under permutations as "families" of codons.

Table III can be extended in an obvious way by analyzing the other three tetrahedral faces successively, in which a guanine base G replaces in turn an adenine A, a cytosine C or a uracil U, so that all four bases are on the same footing. Fig. 16 shows the four-sided tetrahedron laid out flat, so that all of its four faces are contiguous, forming a parallelogram. One should be reminded at this juncture that each intersection point represents *not* the single codons illustrated, but rather the family of codons obtained from it by permutation of its bases in the manner of Table III. In this way we get an interesting interpretation of why sixty-four codons are related to twenty amino acids. While there are sixty-four independent codons, there are just *twenty* families of codons as defined above. This does not mean that codon families correspond in a simple way to individual amino acids, but there is a connection as we shall see.

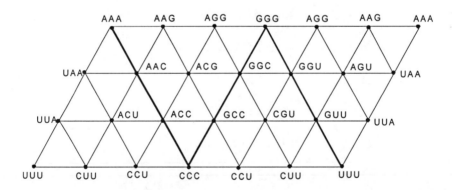

Figure 16. This figure shows how a tetrahedron can be constructed from a plane parallelogram consisting of four equilateral triangles by appropriate folding. (See Appendix C.)

We now turn to the question of how this geometric result might relate in a perspicuous way to the origin of codon degeneracy, to codon bias, and to the origins of the universal genetic code itself. As a first step, we consider

A Physical Approach to Genetic Origins

Fig. 16, which shows how one can construct a regular tetrahedron from an appropriate parallelogram on a flat surface, by successively folding along certain diagonals, e.g. AAA/CCC, CCC/GGG, GGG/UUU (See Appendix C). Note that in this Figure, intersection points are labelled by single codons rather than by codon families as they should be. This is done here for simplicity in labelling only, and the reader is asked to recall that the labelling implies not just the codons shown, but the corresponding families of permutations.

The next step, while not necessary, is remarkably suggestive. Fig. 16 displays the entire set of codon families, and thus the entire set of 64 codons. We now, somewhat ambitiously, try to display the amino acids on the same tetrahedral faces. For convenience, we do this for one tetrahedral face at a time, starting with the face defined by AAA, CCC and UUU, as shown in Fig. 15. Looking first at an apex such as AAA, we observe that this vertex is associated with an angle of 60 degrees. Similarly, for CCC and UUU. In each of these cases, the codon family has only one member (see Table III). From the Universal Table given in Chapter 5, we see that AAA codes for lysine, CCC for proline and UUU for phenylalanine. We use the 60-degree angular spaces to write in these three amino acids in order to enhance the association of codons and amino acids. Turning again to Fig. 15, all other family points (except the one at the center which is the apex for six distinct triangles) correspond to the apexes of three distinct triangles simultaneously. In each case, the codon family has three members, and thus codes for three distinct amino acids. For example, the family of points AAU, has family members AAU, AUA and UAA which code for the amino acids asparagine (asn), isoleucine (ile) and the STOP signal, respectively. For the interior point of the large equilateral triangle in Fig. 15 we see that it is the vertex for six 60-degree triangles. But this point also represents the six member family of codons AUC, ACU, CAU, CUA, UAC and UCA, which code for the six distinct amino acids isoleucine, threonine, histine, leucine, tyrosine and serine. Thus, for every family of codons in Fig. 15, we have the appropriate number of 60 degree triangles to accommodate filling in the corresponding amino acids. This is shown in Fig. 17.

The association between codon families and amino acids carried out for the face AAA/CCC/UUU can be carried out for the other three tetrahedral faces, as well. In assigning amino acids to particular 60-degree angles, where choice is involved, it is more perspicuous to assign them in such a way that amino acids of a certain kind tend to cluster in islands on the tetrahedral face. The result is shown in Fig. 18 which exposes all the faces of the tetrahedron simultaneously, so that all islands corresponding to the various amino acids are displayed together. The result is very satisfying and perspicuous as we shall now discuss.

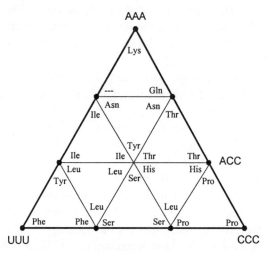

Figure 17. This figure shows how Fig. 15 can be filled in by and inscribed with each amino acid into the 60 degree angular wedges of the figure, in such a way as to associate an amino acid with its codons.

Let us suppose, as argued in Sections 8.1 and 8.2, that thermodynamic considerations would favor the formation of primordial DNA as polymerized repeating base triplets on either strand. The work of Rowe and Trainor [76] suggests that the favored repeating polymers were probably the sequences GAUGAUGAUGAU... and GCUGCUGCUGCU...

For a triplet code, and depending on reading frame, reading from left to right, one could have for the first sequence:

GAU GAU GAU .... in frame 1

G AUG AUG AUG .... in frame 2

or   GA UGA UGA UGA .... in frame 3

A Physical Approach to Genetic Origins 69

For double-stranded DNA, the complementary strand (reading from left to right again) would have:

$$\text{CUA CUA CUA} \ldots \ldots \ldots \text{ in frame 1}$$

$$\text{C UAC UAC UAC} \ldots \ldots \ldots \text{ in frame 2}$$

or $\quad$ CU ACU ACU ACU .... in frame 3

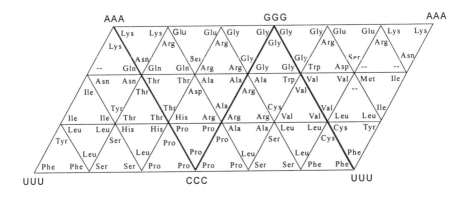

Figure 18. Just as Fig. 17 corresponds to Fig. 15 with the amino acids inscribed corresponding to their codon association, so Fig. 18 completes Fig. 16 by ascribing amino acids in 60 degree angular wedges in association with their respective codons. This can be done in such a way that for each amino acid designation a contiguous island is formed on the tetrahedral surface. It is significant that there are twenty amino acids and twenty families of codons on the tetrahedron.

Thus if we were to allow overlapping reading frames, this one repeating polymer could allow for six different choices of codon GAU, AUG, UGA, CUA, UAC and ACU, which code now, according to Table I, for the five amino acids (asp, met, ile, ser and his) and a stop signal. In fact, overlapping reading frames are known today to occur in primitive forms like phage φX-174 [83]. Methionine seems to play a role in START signals, and the STOP signal is included in this set. Moreover, the other four amino acids cover three of the five functional groups suggested by Dayhoff, Eck and Park [22] on which to fashion a primitive replicating system. Thus this six codon set could by itself code for a primitive replicating system, at least in principle. If one includes the other favored repeating polymer mentioned above, one then has a set of nine amino acids, plus a STOP signal, a set which covers four of the five functional groups suggested in [22]. Moreover, additional codons could eventually evolve, for example by point mutations with consequent changes in reading frames; and by mechanisms in which DNA segments are incorporated into an existing DNA chain.

The general tenor of our argument can now be summarized. Thermodynamic considerations in pre-biotic times would tend to favor the formation of repeating polynucleotide sequences of the following two forms: GAUGAUGAUGAU... and GCUGCUGCUGCU... These sequences are consistent with observations on codon biases of phages in the present era (i.e. that there is a tendency for G to occur in the first slots of codons, A or C to occur in the second slots, and U to occur in the third slots). These observations are suggestive as to how primitive replication systems may have developed, particularly since reverse strand transcription and the use of overlapping frames is known to occur in some primitive systems in the present era. In short, present codon biases in primitive organisms may provide a clue to events in the distant past. This suggestion is strengthened by the results of the cluster analysis carried out in [77] and described in Section 8.2 above.

It is interesting to note from the universal code, Table I of Chapter 5, and from the concept of codon families illustrated in Table III of the present chapter, that the family concept is quite independent of the degeneracy question. This follows from the observation that each codon within a family codes for a distinct amino acid, different from the amino acids coded for by other members of the same family. *This is a significant observation.* It means that at least two point mutations would be required to convert any particular codon to any other codon in the same family, thus indicating the stability of the family concept.

9. Reductionism versus Holism

Reductionism has to do with the philosophy and procedures whereby one attempts to understand things by taking them apart, in so far as this is possible, and reducing their analysis to the study of their natural components and the interactions or interplays among these. Holism is characterized by an attitude that something important is lost in the process of breaking things down into components, and trying to reconstitute the whole in terms of the separate entities. In some sense, one can characterize reductionism and holism as "bottom up" and "top down" approaches to general understanding. A thoughtful discussion and analysis of the relationship of holism to reductionism is given by Charles Lumsden in his essay "Holism and Reduction" [59] in the book *Physical Theory in Biology* [60, pp. 17-44].

Strong arguments can be made that the reductionist viewpoint characteristic of Western philosophy has reaped rich rewards in natural science. The powerful paradigm of modern physics is, perhaps, the prime example, with its very successful approach to the behavior of matter, first reducing it to the study of molecules and atoms, subsequently the study of atoms in terms of electrons and nuclei, then the study of nuclei in terms of their component protons and neutrons, and eventually the decomposition of these particles and the study of their interactions in terms of quarks, which, in principle, account for the structures of the nucleons themselves, as well as the various glue particles (mesons) which describe their interactions. This reductionist path was an eminently successful approach to an understanding of the nature of matter and was accompanied by revolutions in thinking about our basic concepts of space and time as well as matter. Associated with this success were a host of new technologies, with their concomitant social, economic and political impacts, particularly in the communications industry.

Yet it is interesting that the reductionist approach that led so inexorably to the concept of quarks as the elementary "natural components" of matter, also led to the conclusion that individual quarks, cannot themselves be isolated (so-called quark confinement) and studied individually. The reductionist philosophy has nonetheless pushed on to consider other horizons, the most notable one being the proposal that the fundamental elements of matter are not point particles of common notion, but rather entities called "strings." It is too early to say whether strings really exist and provide a useful basis for the description of all matter, or whether they are in turn explicable in terms of even more basic entities, with no end in sight for the reductionist line. These considerations raise fundamental questions as to the relationship between mathematical structures and descriptions of reality.

Another aspect of reductionism was the discovery of quantum mechanics as the apparently fundamental theory describing the behavior of matter in our universe. Despite its reductionist birth, the quantum mechanics of more than one "identical particle" predicts an incredible degree of holism in the behavior of these particles. It seems that all particles in the universe are either fermion (antisocial) or boson (social) particles in their behaviors with respect to one another. In either class, fermions or bosons, the particles are inextricably related in a very holistic way. Without this quantum statistical behavior, there would be no atomic or nuclear structures as we conceive of them today. (This aspect of quantum behavior is discussed in many places, such as in the book by Trainor and Wise [91, pp. 242-243].)

Other areas of natural science have also developed dramatically under the guidance of reductionist approaches to research and understanding, particularly biochemistry and molecular biology. Some of these advances were made possible through the application of reductionist techniques developed in the rise of modern physics and chemistry. Biological structures and functions became explicable at a molecular level. An early example was the realization that proteins, referred to as the molecules of life in Chapter 4, were really specialized polypeptide molecules, that is strings of amino acids held together by peptide bonds. Under physiological conditions these proteins fold automatically into configurations and shapes with a wide range of properties at the molecular level, accounting at once for the remarkable specificities of structure and function endemic to the existence of living organisms. One could now begin to understand the efficacy of hormones, enzymes, neurotransmitters and antibodies in various organic systems.

The other great step in reaching an understanding of the nature of biological organisms and biological processes came from the discovery and its aftermath of the structure of the DNA molecule by Watson and Crick [95] with its implications for understanding protein synthesis in cells, its essential role in genetic inheritance and in the evolution of the species generally. With these discoveries, ancient biological themes such as selective breeding for animal traits, Mendel's theory of inheritance, and Darwin's theory of evolution took on new lives in terms of behavioral descriptions at the molecular level. Darwin's theory, already controversial in many details, received renewed support from molecular biology and contributed to the modern synthesis of neo-Darwinism. Mendel's theory of inheritance became explicable at the molecular level and his units of inheritance were identified as the now ubiquitous gene.

These great revolutions, however, turning biology largely into molecular biology have become rather self-focusing. Mesmerized by great successes in identifying biological molecules and their properties, it became fashionable to regard molecular biology as the only "true" biology; and granting agencies have tended to favor research projects which have reference to molecular biology in their titles. Criticism of this view has grown to substantial proportions [46], though molecular biology continues to be the canonical approach to research and understanding in biology.

Why the opposition to this purely reductionist approach when it has been so fertile in achieving a deeper understanding of biological structure and function at the molecular level? The answer lies in a closer examination of the natures of observation and explanation. There are various levels of understanding, each with its own parameters of observation and description. Each level of description has its own relevance. For example, the discovery of Mendel's inheritance rules and their validation through observations and study on the role of chromosomes preceded the discovery of the molecular basis of genetics. Yet Mendel's advances were made at a relatively holistic level. The reductionist philosophy tends to assume that lower-level descriptions, can in principle be synthesized so as to account for higher-level behavior. A few additional examples will illustrate the limits to this kind of approach, the problem of emergent behavior being a prime factor, with significance for both physics and biology. (See, for a discussion of these points relating to emergence, the essays by Lumsden [59] and Trainor [88], in *Physical Theory in Biology* [60].)

To illustrate this discussion, let us turn to the paradigmatic example of reductionism in modern physics. At a relatively fundamental level, quarks are regarded as the basic building blocks of matter. While few, if any, physicists today would quarrel with this essential belief, one is reminded that before quarks were either inferred or invented, nuclear reactors were built, nuclear bombs exploded, and radioactive tracers used to revolutionize biology and agriculture. Even today, we face the fundamental limitation that individual quarks do not manifest themselves (quark confinement).

In another context, essentially all physicists believe that quantum theory and relativity are at the base of all physical phenomena. Yet landing on the moon by the U.S. astronauts did not generally require an understanding of fundamental physical laws beyond the "classical" dynamical laws of Isaac Newton in the seventeenth century. Einstein's theories of general and special relativity played no role in the success of the enterprise. Quantum mechanics played little direct role either, except in the design and operation of the sophisticated communication systems used in that venture. If quarks are so fundamental, how can we ignore them? Or can we?

The answer has many aspects, but perhaps the most important of these is that observations on a system can occur at different "levels," and even when it seems reasonable to assume that a more fundamental explanation of an observed phenomenon could, in principle, be achieved at a "lower level" (more microscopic, perhaps), it does not mean that it is either prudent or necessary to do so. Each level of explanation has its particular observational parameters and concepts appropriate to that level [59]. Consider the following example: a blacksmith, generally a male, is a skilled artisan who knows his iron: e.g. the fundamental difference between a piece of iron that is highly malleable, and one that is rigid and subject to fracture. The blacksmith would scarcely think of consulting with a physicist, or even a metallurgist, before fashioning a hook or clevis or repairing the moldboard (which turns over the furrow slice in plowing). Yet the physicist and the metallurgist may well be involved in lifetimes of study on the nature of the annealing process, and on the nature and cause of fracturing. In this example, the blacksmith as artisan knows how to anneal iron rods so they can be bent and shaped, alternatively, how to ensure that a particular rod is stiff and brittle. He knows how to achieve these ancient goals by adjusting the cooling rates of the iron from furnace temperatures to ambient temperatures in the shop. Were he to wait until he had a rather thorough understanding of iron at the ionic level, he would scarce fashion many ploughshares in his lifetime.

In a reductionist approach through molecular biology, ancient problems were resolved and ancient mysteries dropped way. Yet problems still exist at a fundamental level. Mae-Wan Ho [42] discusses in some detail the outstanding problem of coherence in a biological organism. This is a feature of the system as a whole which is unlikely to be resolved by a molecule-by-molecule analysis. Presumably some analysis featuring the coordination of many molecules acting in consort will be necessary to achieve understanding. Ho and Saunders [43] and Goodwin and Trainor [35], among others, have emphasized the importance of studying organisms as whole systems. Particular examples are afforded by the treatment of developmental fields in biology as dynamical fields with considerable success by Trainor, Goodwin, Brandts and Totafurno. These attempts are summarized in Part II of [60].

Horticulture is an occupation requiring a great deal of knowledge about how plants grow and multiply, and in what environments. It is a pleasant occupation, and for many people untutored in molecular biology it provides a stimulating and productive pastime or avocation. In what they observe and do, a molecular description of the various life processes is neither essential nor, perhaps, profitable to them. From a molecular point of view, horticulture is a rather holistic enterprise. Clearly holism and reductionism are not in conflict, but correspond to different levels of observation and understanding. In most cases, these two views complement each other.

Another example where holism and reductionism, if narrowly viewed, might appear in conflict, but if widely viewed can be said to complement each other, is in the practice of medicine. Western medicine is primarily based on a reductionist philosophy with the view that since the body is made up of very distinct organs with very distinct functions, a breakdown of good health suggests that some organ or system in the body is malfunctioning. Growth in medical understanding and practice then calls for the development of expertise in the care of each organ or system separately. Research associated with medicine then naturally takes the path of reduction, so that specific diseases have specific causes and specific cures. Diseases tend to be classified as organ- or system-specific, for example kidney diseases, heart diseases, brain diseases, etc. Chinese medicine on the other hand tends to treat the whole system as in disorder. This holistic view is probably more effective in dealing with many chronic problems which affect several systems in concert. The intervention of acupuncture in North America, for example, was at first regarded by the medical establishment as equivalent to something like snake-oil medicine, but with time has become widely accepted as an important adjunct to Western medicine with a substantial physiological basis.

Finally a few words about emergence. Emergence has been discussed elsewhere (e.g. by Lumsden [59] and Trainor in [88]). Generally we speak of emergence as the manifestation of a property of some system at some higher level of observation which, while explicable in principle in terms of lower-level descriptors, is not usually predicted from a study of the system at this lower level, either because such a reduction is too complicated to carry out, or because the phenomenon itself is so unexpected. Examples in physics abound, such as the occurrence of superfluidity in quantum liquids or superconductivity in some metals at low temperatures. Perhaps the outstanding example in biology is the emergence of consciousness in systems of interacting neurons. The new field of nonlinear studies (68) has furnished fertile ground in both physics and biology for studying emergent phenomena. Indeed, all complex systems show nonlinear behavior, including the systems of common consideration throughout the social sciences. All such systems pose formidable mathematical problems and make the transition from any level of observation to possible lower levels of explanation difficult. Nonetheless, new fields in science and mathematics have grown up around these challenges, and popular interest is spawned by such evocative phrases as fractal dimensions, chaotic systems, Feigenbaum numbers and self-organized criticality [6]. For an exciting introduction to these fields, see [34].

The bottom line is that holism and reductionism are not in conflict, but are complementary approaches to the human understanding of natural processes.

10 . Cultural and Material Impacts of Molecular Biology

Whatever else, the molecular approach to biological understanding promises something of a grand unification and synthesis for the study of living organisms. It is now clear that life has a molecular basis and that we have the knowledge of how to tinker with it at that level, but with unknown ultimate consequences. Knowledge, of course, is not in and of itself wisdom. Knowledge has to do with how things are, wisdom has to do with the applications of knowledge. (See R.C. Lewontin [58].)

What is also clear is that there is a vast revolution in progress as a result of the new understandings which have been achieved in molecular biology, coupled with the drive to apply this knowledge to alter living organisms at the genetic level (so-called genetic engineering). There are two basic driving forces in this revolution. The academic component of the revolution is driven by an inseparable mixture of scientific curiosity and a fierce and intense competition for scientific recognition, fame and fortune on the part of scientists and engineers. The non-academic driving force is the marketplace where shareholders and investors sense inviting prospects for economic gain. This is the driving force *sine qua non*.

What is less clear is the balance between the constructive and destructive forces associated with the revolution in molecular biology. Certainly there is an unassessed potential for both the public good and the public evil. Enthusiasts pushing for dramatic applications of the new understanding arising from the molecular approach to biological organization naturally emphasize its promises for human betterment.

Whatever the justifications of our hopes and fears in these regards, one thing is clear, viz. the need to bridge the gap between the experts who have some personal stake in developments and the bulk of society without the scientific training to make informed assessments of what the biological revolution has in store for themselves and society. Enthusiasts naturally emphasize the promise of human betterment, for example, the treatment of human disease, extending life expectancy and achieving increased food production. Such prospects present the rosy side of the revolution in molecular biology. But there is a darker side to the same revolution, perhaps even very dark. And it is doubtful that we possess the wisdom necessary to achieve a balance between the prospects for good and evil inherent in our knowledge about molecular biology and its applications. In our aggressive global culture where efficiency and profit are perceived as the required driving forces for economic prosperity and the common good, concerns about the future results of genetic engineering are pushed aside, or branded with such labels "luddite" and "negative thinker."

But we ignore the dark side of the molecular biology revolution at our common peril. The Brave New World of Lee Silver [82] could end up as a Nightmare World of unimaginable proportions. In the past, approaches to the betterment of the human condition have relied on education and social change. With naivety and a frightening aspect, the new approach is to engineer the betterment of the human condition at the genetic level. The point to be emphasized here is that the consequences of genetic engineering cannot be predicted nor controlled with any certainty. This point is supported by the certainty of sound mathematical analysis as will be discussed in what follows.

A particular problem to be reckoned with is the growing gulf between the natural sciences and the humanities, a gulf given eloquent exposition by the author/scientist C.P. Snow in his book *The Two Cultures* [84] and taken up with informed passion by Edward Wilson in his recent book *Consilience* [100]. Snow has received much criticism for his views, but they still have an audience and an interest after four decades. It is no longer clear that Snow made the most pertinent division of society into two cultures, however. In modern Western Society, other choices of relevant divisions are possible with their ever widening gulfs, for example, the gulf between the excessively rich and the demoralized poor, or the gap between democratic societies and multinational corporate structures. The gap of most concern in our times may well be between the experts in biotechnology promoting personal and corporate agendas and the political decision makers responding to the market pressures and hype associated with these agendas, but with a profound ignorance of the possible material and ethical consequences.

Wilson [100] first argues that a deeper analysis shows that the humanities are ultimately grounded in the natural sciences, but this connection is either ignored or unappreciated by people in the humanities, at our common peril. There is a high degree of consilience between the various natural sciences, but not between the sciences and the humanities where decision-making is usually centered. The gap can be loosely phrased as between those who have knowledge and those who have power.

One missing link in the chain of debate concerning the advantages of genetic engineering, pertinent to our survival, is the ethos we have developed as a species that our destiny lies in trying to understand nature so that we can control it for our own advantages. There are two rather fundamental concerns arising out of this postural stance. One is cultural—the failure to appreciate that our species does not stand outside of nature, but is a part of it, and we must be concerned with the whole symbiosis of living organisms and not just with our short term self-focussed part.

But a second missing link is even more basic since it concerns creating problems which cannot be solved, even in principle. There is a social gap also connected with this problem. This gap is between a rather small elite group, on the one hand, of applied mathematicians, computer scientists and theoretical physicists who have studied complex, nonlinear systems, and on the other hand, the vast bulk of society (including most entrepreneurs, many biologists and genetic engineers) who have had no analytical experience with them.

It has become increasingly clear in recent decades that almost all dynamical systems in nature are complex and nonlinear. A dynamical system is basically a system of whatever kind, that changes as time progresses, and dynamics is the study of the behaviors of such systems. Complexity is the degree to which such systems require a large number of variables or parameters (descriptors) for their exposition, usually with connections between the variables and parameters, so that changes in the value of one of them effects changes in the values of many others. But nonlinearity is the key word. A nonlinear system is basically any system which is not linear (68). Allopathic medicine for example, i.e. standard medical practice, assumes that if a drug of a certain dosage produces such and such an effect, doubling the dosage produces twice the same effect.

In a nonlinear system, however, even a very small change in some parameter can give rise to a very large effect in the behavior of the system. Moreover, nonlinearity is difficult to analyze from a mathematical or computational viewpoint. A very small change in some system parameter can lead in time to an altogether different system behavior. Also disconcerting is the sensitivity to initial conditions. Two nonlinear systems of identical structure starting from two slightly different initial states can diverge from each other in time and display eventually completely different behaviors.

The sensitivity of nonlinear systems to small changes has given rise to the expression "butterfly effect." It derives this terminology from the possibility that a small butterfly in Japan flapping its wings can lead, in principle, to a large weather change in a region of North America a few days later. An exaggeration, perhaps, but a pertinent one. The phrase "linear thinking" is an epithet applied sometimes to people thinking linearly (i.e. naively) when dealing with nonlinear situations.

The consequence is that for a complex, nonlinear system, one cannot in general predict future behavior from present circumstances even if the general structure of the system itself is known. The sensitivity of such systems to parameter changes is legendary. Many examples can be given relating to everyday experience, but perhaps the difficulty with making long range weather predictions is the most familiar. The vagaries of the stock market is another. Practically all chemical reactions are nonlinear in the populations of the various reactants. Turbulence is a common experience for engineers designing airplane wings or motor boat propellers. Since the cells of every organism in biology are little factories driven by hundreds, even thousands, of nonlinear chemical reactions, biology has a high degree of nonlinearity built into its molecular base. Each cell in its own right is a highly nonlinear factory, capable of taking in metabolites and energy supplies and producing products essential to its own existence as well as to its harmonious relationship to the vast number of other cells in the organism.

In the biological example, we now believe that such systems have been shaped and designed over millions of years of development guided by the Darwinian processes of mutational change and survival of the fittest. *Homo sapiens* is no exception, notwithstanding our somewhat exalted position at the top of the dominance hierarchy among species. We seem to be the only species that has a high degree of self awareness and that has developed an elaborate language and communication skill, both factors contributing essentially to cultural change and development. Cultural change itself is a highly nonlinear and complex process involving inheritable influences, one way or another.

But our cleverness is a two-edged sword. We are clever enough to have developed unique and advanced cultural patterns, but not necessarily clever enough to avoid the pitfalls of our own creations. The old adage warns that a little knowledge is a dangerous thing.

The problems with genetic engineering are manifold, but central to most concerns is the unpredictable nature of complex, nonlinear systems, and the havoc one may unwittingly cause by throwing the symbolical wrench into the genetic gears. The present global state of biological organisms is one of relative stability, engineered over countless generations by the natural constraints of mutation and adaption for survival—the balance of nature with new species filling in niches and species generally adjusting to the populations of both their predators and their prey. This relative stability has been achieved by organisms developing corrective mechanisms to natural perturbations or distresses from the environment. In effect, nature has worked its way through the hazards of nonlinear systems by trial and error mechanisms. The trouble with genetic engineering is that the artificial stresses being put on natural systems are not normal, nor has the cellular organization been adapted to them.

But what is genetic engineering? And why is there concern? Genetic engineering has become a generic term signifying the various procedures that have been devised to alter developing and reproducing systems at the genetic level. With modern biotechnology it is a relatively easy matter to alter or disrupt the genome (i.e. the complete genetic makeup) of virtually any living organism, human, animal, plant, fungus, bacterium, virus, anything living. Keeping in mind that each gene, whatever its source, specifies a protein, and the transcription mechanism in the ribosomes of the cells is essentially universal, adding a gene to a genome results in the addition of the corresponding protein to the "natural" molecular melange already in the cell. Can this operation be achieved? Yes it can, by methods discussed in Appendix D. Such operations are called transgenic, if genes from the genome of one species are entered into the genome of another species. See Griffiths [37] for an example which contributed to the identification of DNA as the "genetic stuff."

The consequences of gene transplants are manifested in the new proteins thus added to the normal complement of the cell. A new protein thrown into the "seasoned" milieu of natural proteins in the cell, will inevitably cause changes in the cellular functions. On balance, one expects these changes to be deleterious, as the normal complex and nonlinear interplay of proteins is disrupted. Our knowledge about the behavior of complex nonlinear systems warns us that the resulting behavior of the cell cannot be calculated beforehand. Even if the transplanted gene were introduced for a specific "good purpose," it will in all likelihood cause other difficulties for the cell relating to its normal operation. And to emphasize what has been stated above, the cellular system is a complex of many highly nonlinear chemical processes, so that it is virtually impossible to predict the outcome of the gene transplant, except by trial and error as measured by the response of the individual organism receiving the gene transplant.

Anyone involved in drug design knows about the problem of side effects. The drug is designed to accomplish a certain objective, presumably beneficial, in the cell, but one cannot begin to compute all of the other effects of adding a new chemical to the melange of reactants already present in the cell. One expects side effects but cannot predict them beforehand.

Of course, if "foreign genes" can be added to genomes, "normal" genes can be removed from genomes also. Again the same considerations discussed above apply. One is changing the protein content of the cell, without knowing the full impact of this procedure except through the arduous method of clinical trial and error. (A rough but simple analogue might be adding or subtracting a transistor in a television circuit and observing what happens to the picture). If genetic alteration is in the germ line, then the offspring will tend to inherit the propensity of the cell to adapt to the new protein in its complement. What needs particular emphasis is that once the genetic complement has been altered, there is essentially no way to reverse the procedure. It is easy to add genes to genomes, but very difficult to remove them from a population. Once the "genie" is out of the bottle, there is no way to stuff it back in again.

Any drug designer will have had the experience that the intended effect of the drug is compromised by the appearance of undesirable side effects. Cellular systems are so complex that drugs cannot be fully and reliably tested in the laboratory, but must be tested by trial usage in the field, usually first on animals, then on humans. Even if apparently successful in field testing, one cannot be certain that deleterious side effects will not turn up in special environmental circumstances, or with time delays of days, weeks or months from the time of administration. The tragic thalidomide experience was a case in point. Unexpected side effects led to serious birth defects in thousands of babies even though the intended application concerned only the mothers' states of health.

Genetic engineering, even if and when it achieves the goals for which it is intended, carries with it a multitude of problems, some scientific, some social and many ethical that cannot be predicted or controlled. A few of these are touched on here, but the interested reader can find a modestly comprehensive account of what is in store for us in the recent book, *The Biotech Century*, by the American author Jeremy Rifken [73].

The prime advantage of genetic engineering is touted to be its application to modern medical practice and the elimination of many genetic diseases. To this end, an enormous international effort, with primary impetus coming from academic, governmental and corporate interests in the United States, has been launched to sequence the complete human genome. This means mapping the entire sequence of base pairs on the DNA, identifying the corresponding genes that comprise it, and locating these genes on the human chromosomes.

Once this procedure has been completed (and recent reports suggest that it will essentially be completed in the year 2001), one will then be able, in principle, to tell what the protein contents of the human cells are, and how the protein populations are regulated according to cell type and cell determination. Then it is argued one can begin to deduce the contribution of each gene to the various factors contributing to human behavior, resistance to diseases, growth factors, physical and mental characteristics, etc. Strictly speaking there is no such thing as a genome common to every member of a species. But one can try to establish a "norm" of sorts, and then look for variants from individual to individual in an attempt to identify gene abnormalities that make one susceptible to this or that disease, this or that mental or physical disability, and more generally to assess the gene or gene combinations that contribute to other human attributes, including physical and mental propensities.

One can immediately identify problems arising from the euphoria associated with the prospects of determination of one's genetic makeup. In principle, one's DNA could be used to categorize people as to their probable strengths and weaknesses, leading to automatic discrimination in the market place (such as affecting applications for work and hazards for obtaining insurance). But genome alteration has bigger headaches associated with it. It holds out the possibility of parents having designer children—selecting for size, disease resistance, intelligence, etc. Only the very wealthy in our society could afford such procedures, further contributing to the class society based on connections, influences and incomes. It does not take much stretch of imagination to conceive of military interests in designer soldiers, sailors, and other military personnel, further exaggerating the power of rich nations with respect to poor nations. A sobering reflection on the human genome project is given by Lewontin [58].

Even if science and society were successful in eliminating diseases, promoting fertility and extending the span of life, the cures might be worse than the diseases. At present levels of knowledge and its applications, there are not enough resources to go around, and various physical and social problems are out of control (e.g. the gradual destruction of the protective ozone layer, the increasing desertification of land masses, and global warming). Genetic engineering is bound to exaggerate rather than to solve these problems. Normal social and political change can only attack these problems slowly, but hopefully with some wisdom and forethought. With genetic engineering, we are replacing social progress with mechanical procedures which cannot be assessed and controlled and which threaten the foundations of our civilization. Again the hazards discussed above come to the fore, that genetic engineering may be carried out with good intentions (but don't count on it), however, it cannot be carried out with wisdom, as its consequences are inherently incalculable.

Finally, let us address the technical question of how genetic engineering is accomplished. This is a complicated assignment and we deal briefly with its technical aspects in Appendix D. The basic steps involved are simple, in principle, but require special techniques to carry out. Imagine a method whereby one is able to identify a gene or group of genes in the genome of some organism, copy the identified set and add the copy to the genome of a second "host"organism lacking some desirable attribute, e.g. resistance to a form of cancer. The technical problems involved are challenging, first how to identify a particular gene or set of genes from a donor genome, secondly how to make a faithful copy of it and finally, how to transfer the copy to augment the genome of the host organism. These steps can all be carried out, including genetic cloning of the results if that is desirable, as described in Appendix D. Presumably the intent of such a program is to identify a gene set in the genome of one organization which appears to support some desirable attribute, such as resistance to a disease such as cancer, and transfer the set to another organism which does not have this desirable trait specified in its genome. This is done in the knowledge that the cellular mechanisms for transcribing the added gene sets into proteins in the cells of the receiver organism will operate successfully since these mechanisms are fairly universal in the biological world. The technical procedures are known. The promises and dangers are in the consequences of these procedures.

The other aspect of genetic engineering that is important is to realize that the effects of gene transfers are relatively unimportant in themselves, but their importance is manifested in the products of these transfers, viz. the proteins produced as these are the active agents in biological processes. This realization gives rise to a whole new view of genetic engineering. If the natural protein complement of a cell is altered by additions or subtractions, it is important to try to understand the interactions of the proteins in the cell with the newly added proteins coming from the gene transfer. Apparently this "protein poisoning" is at the basis of mad cow disease [73]. This consideration has given rise to great activity in the field of proteomics. Proteomics is the study of how proteins interact with one another under a variety of ambient conditions. The shapes and forms of proteins (called conformations) and the distributions of electrical charges on their surfaces are complicated to determine through experimentation, and reliance to a large extent has to be placed on theoretical calculations. Such calculations are very difficult to carry out even with the aid of large computing facilities, and resort must be made to modelling techniques and approximation procedures of uncertain reliability. Studies of binary collisions between proteins is already suspect for reliability, and when one considers the complex of thousands of proteins in a single cell, one gets some idea of the difficulties involved in the subject of proteomics.

Whether one actually believes in a God or not, the phrase has crept into popular culture (with some justification) that genetic engineering is somewhat akin to the genetic technicians playing "God", since the techniques are at hand to create entirely new species to the biological spectrum occurring naturally.

To this point in time, organisms have evolved as a result of chance and necessity. Necessity refers to the physical and chemical principles which are always involved in guiding change; chance refers primarily to genetic mutations whose consequences we cannot control, but we can understand and appreciate. From here on in, chance and necessity will be compromised by human intervention. In the past a sense of purpose has been set by the tendency for species to try to survive by taking advantage of adaptions when offered by circumstances. In the future, who will define purposes and how will these be determined?

After all, the human complexity is not determined solely by the number of genes in the genome, but rather by the almost limitless associations and interactions among them.

Glossary

(1) Sir Isaac Newton (1642-1727) was an English mathematician and physicist who revolutionized the basic concepts of physics during his lifetime. On the one hand he proposed an "inverse square" law of universal gravitation, according to which any two bodies in the universe attract each other with a force proportional to the product of their masses and inversely proportional to the square of their distance apart. The proportionality factor is a universal constant applying to any two bodies in the universe. The theory immediately accounted for the motions of the planets about the sun, and the motions of the stars and galaxies throughout the universe. More prosaically, it accounted for our personal weight as an attraction between our mass and the mass of the earth, and for the occurrence of tidal effects on all of the oceans on earth. On the other hand, he developed a set of three equations which described the motions of bodies under any given forces. The result, still largely valid today for most unexceptional circumstances (low velocities relative to the velocity of light) is referred to as Newtonian mechanics.

(2) James Clerk Maxwell (1831-1879) was a Scottish physicist who made fundamental contributions to the thermodynamic theory of atomic and molecular gases, but even more impressively established the basic laws of electromagnetism which united the phenomena of electricity and magnetism. He devised a set of eight differential equations which describe the basic behavior of electromagnetic fields in matter. "Maxwell's Equations" are studied in all serious physics training programs throughout the world and provide the basic theory behind all electrical power systems and all communication systems.

(3) Albert Einstein (1879-1955) is generally regarded as the greatest physicist of the twentieth century, perhaps of all time. He was awarded a Nobel prize in 1921 for his theory of the photoelectric effect, which

clearly established the basic quantization of the electromagnetic field. His theory of Special Relativity was published in 1905 and his General Theory in 1916. The former transformed our view of the concepts of space and time for objects moving with respect to the viewer at velocities close to the velocity of light, which led to the concept of matter as a form of energy. The General Theory further altered our concepts of space and time and led to a new and comprehensive theory of gravitation, still basically extant.

(4) DNA stands for deoxyribonucleic acid, which is essentially the genetic stuff in the chromosomes of cells of all eukaryotic systems, and in virtually all living systems with a few notable exceptions where RNA is used. RNA stands for ribonucleic acid, which in various forms takes part in the essential genetic processes which take place in the cell. The structures and functions of DNA and RNA are discussed in detail in Chapter 5 of this book.

(5) Carolus Linnaeus (1707-1778) was the founder of the modern system of classification for botany and zoology. His system uses a nomenclature consisting of double Latin names, one for the genus and one for the species, e.g. *drosophila melanogaster* is the common fruit fly so valuable in the laboratory for genetic studies.

(6) Georges-Louis Leclerc Comte de Buffon (1707-1788) was a French naturalist who founded the science of paleontology and is noted for his 36 volume compilation of members of the animal kingdom.

(7) Darwinism refers to theories based on analogies to Darwin's theory of evolution in which life forms evolve through small random changes (now identified with mutations) in which competition for survival determines the direction of change.

(8) Rosalind Franklin (1920-1958) and Maurice Wilkins were English biochemists who worked independently, but in the same laboratory (that of J.T. Randall) at King's College, University of London, on the detailed structural analysis of DNA using X-Ray diffraction techniques. They worked in collaboration with Francis Crick and colleagues at the University of Cambridge, who had evolved the theory of X-Ray

diffraction from helical structures. Their work, beginning with diffraction pattern studies on DNA threads pulled from solution, formed the basis of the landmark work of Watson and Crick in deducing the molecular structure of the DNA molecule. Wilkins shared the Nobel prize with Watson and Crick in 1962. Controversy surrounds the question (see [52]) of why Franklin did not share the Nobel prize in 1962 also.

(9) Polypeptides are a class of molecules, including proteins, consisting of amino acids strung together like beads on a string through special bonds called peptide bonds. The nature of these bonds is discussed in Appendix A.

(10) Amino acids form a group of organic compounds characterized by having both a carboxyl group (COOH) and an amino group (NHH) on the same molecule. Their significance for biology stems from the fact that they can be linked in a chain with successive amino acids attached together through the special peptide bond formed from the amino group of one amino acid and the carboxyl group of its neighbor. In general, there is a small subgroup of twenty amino acids that contribute in an essential way to the structure of the proteins that occur in biological organisms. The sequence of amino acids on a string determines the properties of the associated proteins. See text for details.

(11) As discussed in the text, when proteins fold in aqueous solution, the folding pattern is determined by the particular sequence of amino acid in the string. As a result the three dimensional shape or form is characteristic of each protein. The particular protein has a form or conformation which is distinctive in respect to surface features. As a result a specific protein may especially accommodate another because their surface shapes are complementary so that a strong interaction results. Often it is not another protein that has the complementary shape, but a smaller molecule, sometimes referred to as a substrate. When a particular shape or form occurs for a protein which is particularly suited to a chemical reaction, one refers to the active site as a receptor site. Receptor sites are highly specific as to what complementary molecular shapes will be accommodated. When accommodation occurs, reaction rates can be speeded up by many orders of magnitude. In fact, both the

speed and specificity combine to make biological functions of great variety possible.

(12) Transcription is the process of transcribing to messenger RNA the coded information stored on the DNA template for the specification of the amino acid string corresponding to a particular protein. Transcription is accomplished through Watson-Crick pairing between the DNA template and the mRNA strand being polymerized with the assistance of RNA polymerase enzyme.

(13) Translation is the process at ribosomal sites which corresponds to formation of the polypeptide protein chain according to the genetic information carried by the mRNA as it threads through the ribosome. In effect it is a translation from the language of codons on the mRNA to the language of amino acids in the protein chain.

(14) Ribosomes are special organelles composed of protein and a form of RNA called ribosomal RNA abbreviated rRNA. They are made in the nucleus of the cell, and can be found throughout the extra nuclear region, but particularly on the outer surface of the double membrane enclosing the nucleus. Their general structure is known to consist of two units with special functions. The ribosomes are the sites of mRNA translation into protein. As each mRNA codon passes through the ribosomal mechanism, the appropriate amino acid specified by the Universal Code is bonded to the forming end of the protein chain through a peptide bond.

(15) A codon is a sequence of three bases on the mRNA string in the appropriate reading frame coding for a particular amino acid, according to the universal Table I in Chapter 5.

(16) Lipids are essentially fat molecules which can occur in a great variety of compositions but all have a similar construction. They are characterized by a polar "head group" with one or two long hydrocarbon tails. The head groups are hydrophilic, preferring a water environment, while the hydrocarbon tails are hydrophobic, preferring not to be in contact with water. In an aqueous environment, lipids will tend to orient themselves in such a way as to minimize the energy of contact, e.g. by forming

vesicles with the hydrocarbon tails pointing inward and the polar head groups pointing outward into the aqueous medium. A special importance in biology comes from their tendency to form lipid bilayers (membranes) separating two aqueous regions.

(17) Polysaccharides are carbohydrates which include sugars, starches, cellulose and insulin, which consist of long chains of monosaccharides. Monosaccharides are relatively small sugar molecules which cannot be hydrolyzed to form even simpler sugars. Examples are glucose, lactose and fructose.

(18) Differentiation refers to the general process through which cells in the multicellular system become different in structure and therefore function. Beginning with the zygote (the fertilized cell) through cellular division and differentiation, various kinds of cells are produced, e.g. red blood cells, lymphocytes, muscle cells, neurons, etc., each designed to serve some specific purpose, such as growing hairs, excreting hormones, etc.

(19) Entropy is a thermodynamic quantity which essentially measures the number of states available to a system at a given temperature and density. The second law of thermodynamics is equivalent to the statement that the entropy of the universe is increasing, because in all thermodynamic changes, the number of available states either remains constant or increases.

(20) The second law of thermodynamics has many equivalent formulations. One is that, in any physical or chemical change, the entropy of a system either remains the same (reversible changes) or increases (irreversible changes). The most prosaic definition is that heat always flows from a hotter region to one which is cooler, and in such a heat flow, there is an upper limit to the fraction of heat flowing that can be used to do useful work.

(21) From glossary item (20) above, it is clear that to obtain useful work from a source of heat, one must have a cold reservoir to which heat can flow from the heat source.

(22) Electromagnetic waves are oscillating electric and magnetic fields moving through a medium, or even a vacuum. The quantum particles in such fields are called photons. Electromagnetic waves manifest themselves in a number of forms depending on the frequency of oscillation in such a way that the higher the frequency the more energetic are the constituent photons. Proceeding from low to high frequency one has radio waves, radar waves, infrared waves (radiant heat waves), optical waves, ultraviolet waves, X-rays, gamma rays, etc, High frequencies correspond to short wave lengths, low frequencies to long wave lengths. To probe objects like molecules that have small sizes, one needs very short wave lengths. X-rays are ideal because they can be easily produced and have wavelengths the order of molecular sizes. The smaller the molecule, the shorter the wave length necessary to photograph them. When X-rays are scattered off molecular structures, the resulting patterns of the scattered photons reveal something about the molecular structure [101].

(23) Linus Pauling was an American chemist who received the Nobel prize for Chemistry in 1954 for his work on molecular structure and the nature of the chemical bond. He was particularly distinguished by receipt of a second Nobel prize in 1962, this time for peace. He had been involved with attempts to prevent testing of nuclear bombs because radiation from fallout threatened public health.

(24) The transcription process from DNA to RNA is not a simple process. The RNA is transcribed piecewise and the pieces assembled subsequently with the aid of enzymes in the protoplasm.

(25) The French scientists Jacob and Monod shared the 1967 Nobel prize with Andre Lwoff. Jacob was primarily a geneticist, Monod a biochemist. Their discovery of the *lac operon* set a new vision of how genes are involved in cellular regulation processes. Monod's book, *le Hazard et la Necessité* (Chance and Necessity) was a great favorite in Europe and America in extending to the general public a view of how the physical laws and chance combine in the explanation of life phenomena.

Glossary

(26) In the succession of bases on either strand of the DNA backbone, the sugar groups (2-deoxyribose) are bridged by a special bond called the phosphate bond. The phosphate atom is bonded simultaneously to four oxygen atoms. Phosphate bonds are discussed in item (58) of this glossary in connection with a discussion of the energy carrier ATP (adenosine triphosphate).

(27) A base in the genetic sense is a molecular unit of either the purine or the pyrimidine type, the former characterized in structure by two ring groups each containing two nitrogen atoms, the latter by a single ring group also containing two nitrogen atoms. Cytosine and guanine each carries additionally an NHH group which contributes an extra hydrogen bond to CG pairing as compared to AT pairing (see Fig. 4). For the DNA structure of biological organisms, the purines are adenine or guanine (A and G), the pyrimidines either thymine or cytosine (T and C). The Watson-Crick pairing rules require A to bind with T, and C with G as discussed in glossary item (29).

(28) There are a variety of different modes of chemical bonding. In hydrogen bonding a hydrogen atom is shared between two atoms, these being referred to as donor and receptor, according to which atom the hydrogen is primarily associated. The primary protein structure is the sequence of amino acids bonded together by peptide bonds. In the protein folding process, however, hydrogen bonding plays a crucial role, bridging between adjacent amino acids on different sections of the protein string. Hydrogen bonds are much weaker than ionic bonds which occur as a result of the attraction of charges, but are considerably stronger than Van der Waals bonds characteristic of the bonding between neutral atoms or molecules.

(29) We refer in this book to the AT and CG pairing elucidated by Watson and Crick [95] as "Watson-Crick pairing."

(30) Marshall Nirenberg's laboratory was located at the National Institutes of Health in Bethesda, Maryland, USA. Francis Crick had his laboratory at the Medical Research Council Laboratories at Cambridge, UK.

(31) Polymerization is a process by which molecular groups (often identical or related) are brought together in a string or chain-like fashion In many cases polymerization requires the assistance of enzymes. A notable example is DNA polymerase which synthesizes a chain of nucleotides in the presence of a template directing Watson-Crick pairing. This enzyme is so efficient that mistakes due to incorrect pairing are about one in a million base pairs.

(32) When a cell divides in general (there are always some exceptions in biology) the daughter cells possess the same chromosome structure and hence DNA complement as the parent cell. This duplication process, which depends on Watson-Crick pairing, is *replication*.

(33) The nature of the genetic code is discussed in detail in Chapters 5 and 6. Clearly, in any succession of bases on a section of DNA or mRNA, the codons will be different depending upon the starting point, since codons consist of three successive bases. In other words, one has to be in the correct reading frame to get the genetic message correctly. In some bacteria, two reading frames are utilized over some sequence of bases so that a different genetic message is read from each frame, but with coherent results. Such overlapping frames do not occur in eukaryote DNA's.

(34) Organelles are small organs in the nuclear or cytoplasmic volumes which form part of the cellular structure and which carry out specific functions, e.g. the ribosomes (glossary item (14)) and the Golgi apparatus, a system of folded membranes in the nucleus involved in the sorting and packaging of glycoproteins.

(35) Ribosomal RNA is still another form of RNA, denoted rRNA, which with protein forms the structure of the ribosomes where the translation process takes place.

(36) Adaptor molecules are key molecules in the translation process whereby the codons on mRNA direct the construction of protein by choosing the correct amino acid to add to the growing protein string. The adaptor molecules are tRNA molecules (Fig. 7) with a specific anticodon as part of the structure and with the corresponding amino acid attached at one

Glossary 95

of the free ends. When the anticodon matches the codon of the RNA string which is in the active site on the ribosome, the amino acid is attached to the forming end of the protein chain by a peptide bond.

(37) A codon as discussed in the text is a unit of genetic information—three successive bases in the correct reading frame corresponds to one of the twenty amino acids in the protein chain. An anticodon is a set of three bases obtained by complementary pairing with a codon.

(38) pH is a measure of the relative acidity of an aqueous solution. It is in fact a measure of the concentration of dissociated hydrogen ions in solution. A pH of 7 indicates a neutral solution, a pH less than 7 an acidic solution, and a pH greater than 7 an alkaline solution.

(39) Some strings of amino acids, taking advantage of the rotational features of the peptide bond, find their energy minimum by coiling into a helix, the so-called alpha helix. Successive coils are stabilized by hydrogen bonds between them. In some protein structures (globular proteins), alpha helical coils and beta pleated sheets both occur.

(40) Proteins are strings of amino acids joined together through peptide bonds. This structure is referred to as the primary structure of the protein. In aqueous solution the protein folds to form various secondary or tertiary structures. A common structure is the alpha helical coil made possible by the rigidity of the peptide bond itself but with relatively free rotation of the amino acid structure on either side. Another prominent structure is the beta pleated sheet in which the protein string weaves back and forth to form "sheets," which are then tied together by hydrogen bonds.

(41) Microtubules are cylindrical molecular structures made from two kinds of small proteins, alpha and beta tubulin, in appropriate arrangement. Microtubules form important structural elements in the cellular architecture but also contribute to a variety of cellular processes by virtue of the fact that they respond by growing or decaying at their open ends depending upon the salinity of the cytoplasm.

(42) Exchange forces are a purely quantum mechanical phenomenon in which neighboring atoms share an electron with the overall effect of causing an attraction between them.

(43) The inverse square law applied to gravitational forces states that the force between any two objects falls off in proportion to the square of the distance between them. This law also applies to the form of interaction between two electrical charges.

(44) In the fertilization process in eukaryotes the offspring share genes from both parents, but the sharing process is not unique, so that different offspring from the same parents usually differ remarkably from each other in their genetic complement. This then adds to the variability of the species which increases with each generation.

(45) Gregor Mendel (182 1884) was a Moravian monk who discovered the basic laws of inheritance from studying the hybrids formed from crossing different species of peas, e.g. peas that were yellow or green, wrinkled or smooth, etc. The importance of his work went almost unnoticed until the advent of modern molecular biochemistry, where detailed structural determinants could be studied.

(46) Sol Spiegelman carried out important experiments at the University of Illinois during the period 1949 to 1969 bearing on the origin of life. These experiments demonstrated the remarkable ability of viral RNA to replicate under favorable conditions. The self catalyzing properties of RNA strengthen the plausibility of an "RNA world" at the origin of life. A simple overview of these experiments and the achievement of a "slimmed down" replicator dubbed "Spiegelman's monster" are given in [20, p. 127].

(47) Mitosis is the process whereby a somatic (body) cell divides into two daughter cells each possessing the full DNA complement of the parent cell. A mitotic cell is one undergoing division through several well-defined stages including the stages during which chromosome and DNA duplication take place, followed by actual division.

Glossary

(48) Meiosis is a special type of cell division resulting in daughter cells with half the chromosome number of the parent cell, preparatory to the fertilization process whereby sperm penetration of the ovum sets in motion a number of structural changes in the ovum, including the crossing over process whereby the new cell of the offspring (zygote) acquires chromosomal contributions from both male and female gametes. The whole process is complicated, and there are many species-specific variations in the process. A reasonably simple explanation of these processes is given in [11].

(49) The common microscope is a light microscope, but its resolving power is limited by the relatively long wave lengths of visible light. When it was discovered that electrons (as quantum particles) have wave properties and that these waves diffract as one would expect of a wave phenomenon, the concept of an electron microscope was born. For electrons of modest kinetic energy, the associated (deBroglie) waves have a much shorter wave length even than X-rays, thus allowing greater resolution for examining molecular structures.

(50) A phage is a microbial virus which is parasitic in the sense of infecting the bacterium in such a way as to take over the reproductive machinery of the cell. The virus itself does not have its own reproductive machinery, but it uses the bacterium machinery to replicate itself many times, eventually bursting the bacterial wall and spreading its progeny to repeat the same scenario on other bacterial cells.

(51) A mutant is the product of an inheritable genetic alteration, which may e.g. be caused by radioactivity or other agents (called mutagens) introducing a change in one or more genes in the genome. Mutants correspond to organisms in which one or more genes are altered. The altered genes are called alleles and have different forms from the normal (wild type) genes.

(52) Crossing over is the exchange of genes between homologous chromosomes, one from each parent resulting in progeny, which on development show inherited characteristics of both parents.

(53) Recombination is the rearrangement of genes on chromosomes by the crossing over process to form a combination different from either parent.

(54) A codon is a set of three successive bases in any particular reading frame coding for a specific amino acid. In principle, a genetic system could use another reading frame as well so that any base would contribute to more than one codon. Crick [17] showed that in fact the genetic code was non-overlapping. Some viruses show an exception to this rule.

(55) Degeneracy refers to the fact that more than one codon can code for the same amino acid. For example, the amino acid serine is coded for by six different codons. This presents a problem only in the fact that codons and amino acids are in a many-to-one correspondence rather than in a one-to-one correspondence.

(56) Mitochondria are special organelles which form the energy batteries for many biological functions, including muscle actions and endothermic chemical reactions which could not proceed without their assistance. In mitochondria pyruvate is transformed into co-enzyme A. In the citric acid (or Kreb's cycle) which also takes place in mitochondria, twelve high energy phosphate bonds are generated from each two carbon fragment, reducing them to water and carbon dioxide in the process. (See Stryer [87] for details.) In aerobic systems, glycolysis in the protoplasm converts glucose to pyruvate with the production of ATP. The pyruvate is an input to the citric acid cycle in the mitochondria.

(57) A prokaryote is a single-celled organism which has neither a distinct nucleus enclosed in a membrane, nor other specialized organelles. Examples are bacteria or blue-green algae. Eukaryotic cells have a central nucleus.

(58) Adenosine triphosphate is the principal energy carrier in eukaryotic systems by virtue of its high energy phosphate bond which gives up this energy on hydrolysis to adenosine diphosphate. It is produced in the mitochondria (glossary item 56) through respiration and oxidative phosphorylation. Mechanical movements and endothermic chemical

reactions depend on the conversion of ATP to ADP for their energy supply. Energy transfers in biosystems are very efficient (see Ho [42]). According to Stryer [87], a resting human utilizes about 40 kgm of ATP in a twenty four hour period.

(59) ATP has 3 phosphate bonds. Each bond involves a phosphate atom bonding to three oxygen atoms (see Fig. 11).

(60) The Krebs cycle is also called the citric acid cycle. In effect, it involves conversion of pyruvate in a series of enzyme assisted reactions to produce 12 high energy phosphate bonds from each two carbon fragment which fully reduce to water and carbon dioxide. The cycle which is complicated is described in detail in Stryer [87].

(61) Self-organization refers to a system which spontaneously organizes its parts into a coherent whole. In many instances this requires a steady throughput of energy to maintain it far from thermodynamic equilibrium (see Prigogine and Stenger [70]).

(62) Archaea represent possibly the oldest forms of living organisms, being the most ancient of the three domains of primitive life forms, archaea, bacteria and eukarya. Davies [20, Chapter 7] discusses theories on the origin of life and the ancient separation of archaea from bacteria about four billion years ago.

(63) Superbugs is a term sometimes used to describe bacteria which have become resistant to antibiotics. But in the present context, it refers to bacterial organisms which thrive under unusual environmental conditions, e.g. the thermophiles which thrive at temperatures much higher than normal physiologic temperatures, or bacteria which live at great depths under the land surface and digest minerals for a living. (See [20, Chapter 7] for an excellent discussion of the discoveries of these superbugs.)

(64) There are regions on DNA, called introns, which do not code for amino acids in the translation process. Regions which do are referred to as exons. Introns and exons are often interspersed in regions of DNA. The role of the introns is not known at the present time, but it is generally

assumed that they play some higher level function in the overall information system in the genome.

(65) Molecular dynamics refers to the interaction of molecules in cells and the changes that take place in their structures and functions with the passage of time. The competition between molecules for survival is referred to as molecular Darwinism.

(66) Ising was a physicist who in 1928 proposed a simple model of ferromagnetism in which the spins of neighboring atoms interact in such a way as to lower the system energy when the spins are parallel. Simple in construct, it failed to provide a correct theory of magnetism, but it provided a useful model in application to a variety of other problems.

(67) Periodic polymers are polymers with repeating units. The units themselves may be simple or complicated molecular structures.

(68) A linear system is one in which the significant descriptors bear a proportional relationship to one another. If one descriptor (parameter) is doubled, a related descriptor is either doubled also or possibly halved. A non-linear system is one in which such relationships are not linear.

Appendix A. Peptide Bonds and Protein Structure

Amino acids strung together in a chain with peptide bonds between them form what is referred to as a polypeptide. Polypeptides differ in terms of the specific sequence of amino acids in the chain and in its overall length, the presence or absence of branches, etc. Under appropriate conditions, the linear chain folds up on itself to form a more-or-less specific three dimensional structure. On increasing the temperature appropriately, the chain unfolds, the process being referred to as denaturing. The amino acids themselves, although possessing some common structure, vary from one kind to another in terms of the so-called residues or side chains designated as R in chemical diagrams. The basic structure of an amino acid is shown in Fig. 9 of the text. This figure also shows how a peptide bond is formed between two amino acids. The so-called alpha carbon in the amino acid shares one of its four bonds with a hydrogen atom, one with the residue R, one with the amine group NHH and one with the carboxyl group COOH. (The structure is somewhat altered when the amino acid is in solution and takes on a polar form.) As indicated in Fig. 9, the carboxyl group on one amino acid and the amine group on another can interact to form a peptide bond with the release of a molecule of water. Formation of peptide bonds requires an energy supply mechanism and is usually facilitated by the presence of appropriate enzymes. The bond itself is in the form of a rather flat plate, (planar structure) with rotatable bonds to the adjoining amino acids. This structure assists in the folding process by virtue of the rotatable bonds. Strings of amino acids joined one to another through peptide bonds are referred to as polypeptides.

On formation of a peptide bond, electrical neutrality is achieved in the bond, so that any excess positive or negative charge if present is carried by the residue on the amino acid itself. It is the residues that give polypeptides their distinctive features. Some residues are polarized, others are not. Residues differ in size and charge distribution, but can be roughly categorized into three groups: hydrophilic, hydrophobic and neutral depending on their interactions with a water medium, the hydrophilic amino acids preferring contact with water, the hydrophobic preferring no water contact, and the neutral being indifferent to the presence of water. These features of amino acids contribute to the overall shape or form of a folded polypeptide. In globular proteins, for example, the hydrophobic amino acids tend to be located on the inside of the molecule, away from contact with the aqueous milieu, whereas the hydrophilic amino acids tend to be on the outside of the molecule interfacing with the aqueous milieu.

Proteins are a special subclass of all polypeptides which have been selected in evolution to perform special structural and functional tasks. Almost universally in the biological world, proteins are polypeptides constituted from a pool of just twenty amino acids. (These are listed by name and associated RNA codons in Table I, Chapter 5).

Even so, there are almost countless proteins occurring in biological organisms, selected in individual organisms and in individual cells of each organism, according to the structural and functional properties required. A single human cell, for example, can depend upon literally thousands of different proteins for its proper functioning.

One can ask the difficult question, out of the vast number of polypeptides possible, how has nature come to select the subclass of proteins? What is it about proteins that makes them special polypeptides? The answer probably lies in selection at a molecular level by a Darwinian process of survival of the fittest. For a given structural or functional process, those molecules which performed best win out in the competition for survival of cellular performance and eventually of organism performance.

But what were the special properties of proteins that lead to their survival in different life forms? Some clues come from analysis of protein folding as nonlinear dynamical systems (Jian Min Yuan, organizer, Mini-Symposium on Complex Dynamics of Proteins, SIAM Pacific Rim Conference, Maui, [50]). For biological proteins, the shape of the folded protein and the speed of folding are essential features. Protein shapes are a part of efficient protein activities, especially so for enzymes, antibodies and neurotransmitters, where receptor site specificity is key to efficiency and to speed. Speed is an essential factor in biochemical reactions since the interaction of large biomolecules is involved with assistance from specific enzymes. There is evidence that proteins are remarkable for the speed with which they fold from a denatured state thus enhancing their biological effectiveness. Moreover, their folded forms have high specificity in terms of overall shape

Appendix A

and form, so that their utilization is reliable from application to application at the molecular level. The challenge to the theorist in biology is very great in coming to an understanding of both of these features, the speed and reliability of folding.

One of the primary folding units found in many globular protein structures is the alpha helix. It is of particular significance that these helices are all right-handed, whereas left-handed helices are equally likely on a probability basis. In fact, helical coils of amino acids made artificially show equal probabilities of right and left helical forms. This observation underlies one of the most convincing reasons to suppose that life had a unique origin. If life originated in many separated events, one would expect an equal distribution of right and left helical proteins.

Of historical interest is the fact that the first protein to be sequenced for its amino acid structure, which established a linear correspondence with the nucleic structure coding for it, was carried out by Sanger [79, 80] on bovine insulin. Ingram [47] was the first to confirm a molecular disease by showing that hemoglobin for patients with sickle-cell anemia differed from their normal counterpart by a single amino acid error in the long string of amino acids constituting this protein. This example, as well as many others which have since been found to underlie genetic disease, shows how sensitive a protein may be to a single amino acid defect. In sickle-cell anemia, this one error which affects protein folding and protein behavior has devastating observable macroscopic effects. The resulting red cells abandon their characteristically circular forms for the telltale crescent forms which identify the disease in blood samples under the microscope.

Appendix B. Procedure for Introducing a Metric Free Distribution of Points in Codon Space

In [76] in a refined analysis, a metric free distribution of possible points in codon space was introduced in anticipation of a cluster analysis of known DNA sequences. For a specific DNA sequence of length N, the numbers $a(1)$, $t(1)$ ...up to $g(3)$ in Chapter 8, equation 8.2 lead to a specific point in codon space. However, many other sequences could, in principle, share the same point in codon space. In fact, a little reflection shows that this number is

$$\frac{N!}{N(T)!N(A)!N(G)!N(C)!}$$

where $N(T)$ is the number of T's occupying slots in the sequence. Also the functions $N!$ are just the usual factorial functions defined by $N! = N(N-1)(N-2) \ldots (1)$. One can then calculate a probability function in codon space for a sequence of length N and compare the probability of any given sequence to a random distribution in order to detect anomalies, if they exist.

Having plotted many sequences as points in the nine-dimensional codon space, it is interesting to see whether any clustering of points occurs, and if so, with what significance. In [76], 338 sequences were plotted from the Genbank Library at Los Alamos corresponding to a wide range of protein genes, ribosomal RNA, transfer RNA and even non-coding regions of DNA. Clustering was clearly evident but difficult to assess in the distribution of points in the nine-dimensional codon space. A non-hierarchical cluster analysis devised by Anderberg [4] in another context, was used to organize the points into natural non-overlapping clusters. The algorithm carried out in four steps was as follows:

1. Choose N initial points to define starting centroids of clusters. ($N = 16$ was chosen).

2. Assign each datum point to the nearest cluster.
3. Calculate N new centroids as average positions in each of the N clusters.
4. Repeat steps 1 to 3 until convergence (i.e. until no new data are reassigned in step 2).

Appendix C. Fun with Tetrahedra

Consider the parallelogram constructed from four contiguous equilateral (60 degree) triangles arranged successively with base down, base up, base down, base up, as illustrated in Fig. 16 of the text. Cut the parallelogram out from a sheet of construction paper and fold this planar cutout along the three common sides between successive triangles (shown in heavy lines in Fig. 16) until the four triangles form a closed tetrahedron (with no open seams). The four sides are, of course, identical. Choosing anyone side as a base, the tetrahedron can be regarded as a prism with three equilateral sides and a distinct apex. We refer to this tetrahedron as the basic tetrahedron in what follows.

Since these basic tetrahedra are easy to make, it is instructive to make several of them (at least four or more) and use them as building blocks in an attempt to construct a larger tetrahedron with, say, two or three times the dimension of the original basic tetrahedra. With two times it is not possible, but it is a challenging exercise with interesting results, nonetheless. One can regard the exercise as a three-dimensional tiling problem in a tetrahedral space.

Appendix D. Genetic Engineering

What is genetic engineering? Like all engineering it is the application to human purposes of the knowledge gained from fundamental science, in this case genetic science, a child of biochemistry. Applications involve the movement of genes within species, between species, and between unrelated organisms, with consequent modifications of the host genome possibly affecting the structure, growth and behavior of the host organism. Genetic engineering includes the removal of genes from the genome, the addition of genes to the genome, and the transfer of genes from a donor to a host.

One should keep in mind that the genes are only a part of the information system; it is the gene products, the proteins, that are affected and which may alter the structure, form and properties of the host organism. In eukaryotes, most cell types have nuclei with the complete DNA complement carried on the nuclear chromosomes of each cell. Changes in the genome of the germ cells lead to modifications in the protein complement of each cell which affects cellular differentiation and determination, and ultimately alters the form and behavior of the adult organism. Genetic modification cell by cell is virtually impossible, but genetic modification of cells in the germ line are passed on to all nucleated cells in the developing organism. With modern biochemical techniques it is relatively easy to modify the genome, but it is difficult and generally impossible to reverse this process and restore the original genome once breeding and development have taken place.

It is estimated that the human genome has about 100,000 genes, which have now been identified in a preliminary stage (year 2001) and the function determined for perhaps 5000 to 10,000 of these. The Human Genome Project to identify all genes in the human genome is expected to be completed and verified by the year 2005 (see for example, Jaroff [49]).[6]

---

[6] These numbers will, no doubt, be altered as the genome analysis proceeds. An interesting report on preliminary drafts by the two competing groups working on the human genome is given by Julie Karow, "Reading the Book of Life." *Scientific American: Explore!* February 12, 2001. Both groups (the international consortium of academics referred to as the Human Genome Project and Celera Dynamics, a private U.S. group) agree on a major surprise, that

The center piece of genetic engineering is recombinant DNA technology. This technology allows DNA fragments from one source to be incorporated into the DNA of a second (host) genome. The technology involves fragmentation of DNA from one source, selecting a desired fragment (containing one or several genes) and inserting it into the DNA of another organism, thus modifying its genetic makeup and its information content. A detailed discussion of these techniques is given in many places, e.g. in Kornberg [55, pp. 672-679] and in Friefelder [30, Chapter 20]. Here we discuss the matter in broad outline only.

At the basis of recombinant DNA technology is the universality of the genetic code, and the more limited universality of the cellular procedures for reading the code and transcribing it into amino acid language with consequent protein assembly. This enables genes from different organisms to be added to the host genome and to be expressed by the genetic machinery of that host.

Recombinant DNA surgery can be divided into several stages. The first stage depends critically on the existence of the so-called restriction enzymes (nucleases), of which there are many kinds with specific properties of fragmenting a DNA segment by cutting the duplex structure at specific points along its length. A particular fragment may carry several genes, which can be selected according to the particular purpose in mind.

The next step is to find a vehicle (referred to as a vector) to transport the fragment of DNA for insertion into another (host) genome at some later time. The usual vectors are plasmids or phages, the former relatively small, circular pieces of DNA found in bacteria, the latter a suitable phage (bacterial virus). The same restriction enzyme that was used to create the fragment from the original (donor) DNA, is now used to cut the plasmid or viral DNA so that one has some assurance that the donor DNA fragment can be inserted and incorporated into the plasma or phage DNA as the case may be, thus modifying their respective genomes. An enzyme which assists this process is called a ligase.

The next step is to clone the fragment DNA in order to obtain workable amounts of the fragment. This is done by inserting the plasmid into a bacterium in a natural way, or in the case of a phage, infecting the bacterium so that the modified phage DNA now incorporated into the bacterial DNA is

---

the human genome contains substantially fewer genes than had been anticipated. However, the margin of error in identification and interpretation is still quite large, and final results await further clarification.

reproduced (cloned) many times as the bacterial colony grows in a normal way. Thus, from one or a few identical fragments, one is able to obtain milligram amounts of the fragment DNA. The fragments can be recovered from the bacterial DNA by cutting them out using the same restriction enzyme that was used to produce them in the first place.

Applications of this technology range widely. There are medical applications where defective genes can be replaced by normal genes in a germ line. But this technology also helps to identify which genes are defective in the first place. It also assists in the study of the protein complement of cells, and how this is altered by the output of defective genes. These are applications which in some sense may be regarded as socially "good." But there are other applications already extant which are more of a mixed bag ethically. The alteration of animal and plant genes to make the animals or plants grow more rapidly or to achieve larger sizes may have hidden consequences where the modification of the gene complement leads to unforeseen and unwanted consequences.

One contentious application now posing problems is the genetic alteration of the seed used in field crops to make them resistant to certain chemicals. The strategy is this. Weeds in competition with the cash crop are an expensive factor to contend with. Usually this situation is handled by spraying the crops to kill the weeds. However, the chemical sprays also do some damage to the cash crop as well. Solution: genetically alter the seed so that the crop has a resistance to the chemical sprays being used on the weeds. This is certainly a great solution for the farm company that supplies both the spray and the genetically altered seed which has now been patented. Thus the company not only is free to continue selling the chemical spray, but it also has control of genetically altered seed including its price to the farmer. Farmers are enticed to join the program by the promise of greater yield for their cash crops, but once they join they have to stay committed. There is evidence that some seed that was not genetically altered, nonetheless has been infected with the modified genes by natural gene transfer mechanisms, so the patents are infringed unavoidably.

The applications and associated technological improvements of recombinant DNA technology seem endless. We end this Appendix by illustrating developments in this field by reference to two articles published in Scientific American. In one Robert Miller [64, p. 67], describes how bacteria swap genes in nature by any of three processes, transduction, transformation and conjugation (bacterial sex). He discusses these processes and how they work in order to better assess the risks in genetic engineering that altered bacteria of harmful variety might enter into the environment. In the other Haseltine [39, p. 92] discusses the discovery of genes for the new medicine. The central point of this paper is the use of mRNA rather than DNA as a starting point in recombinant DNA technology. The point is this. In eukaryotic genomes, a large fraction of the DNA is not eventually transcribed (confer discussion of introns and exons in glossary item (64)), whereas the mRNA which is processed after transcription, is all translated. The trick then is to start with mRNA, and construct a modified double stranded DNA from it using the enzyme reverse transcriptase. The modified DNA produced then corresponds to the expressed portion of the original DNA, without the portions that are not transcribed in any case. This modified DNA can then be fragmented in the usual fashion with restriction enzymes and one now has the knowledge that the fragments correspond entirely to genes. This simplifies the whole process described above for inserting genes from the donor DNA to the host DNA.

References

[1] Alberti, S. "Evaluation of the Genetic Code, Protein Synthesis and Nucleic Acid Replication,"*Cellular and Molecular Life Sciences.* 1999, **56**, 85.
[2] Alberts, B., Bray, D., Lewis, J., Raff, M., Roberts, K. and Watosn, J.D. *Molecular Biology of the Cell.* New York: Garland Publishing, 1983.
[3] Alberts, B. Bray, D., Lewis, J., Raff, M., Roberts, K. and Watson, J.D. *Molecular Biology of the Cell*, 3rd ed. New York: Garland Publishing, 1994.
[4] Anderberg, M.R. *Cluster Analysis for Applications.* New York: Academic Press, 1973.
[5] Avery, O., Mc Leod, C.M. and McCarty, M. "Studies in the Chemical Nature of the Substances Inducing Transformation of Pneumococcal Types. Induction of Transformation of a Dioxyribonucleic Acid Fraction Isolated from Pneumococcus Type III," *J. Expt. Med.* 1976, **79**, 137.
[6] Bak, P. and Chen, K. "Self-Organized Criticality," San Francisco: *Scientific American*, January 1991, pp. 46-54.
[7] Barricelli, N.A. "On the Origin and Evolution of the Genetic Code, I. Wobbling and its Potential Significance," *J. Theor. Biol.*, 1977, **67**, 109.
[8] Beadle, G.W. and Tatum, E. "Genetic Control of Biochemical Reactions," *Neurospora, Proc. Nat. Acad. Sci.,* USA, 1941, **27**, 499.
[9] Behe, M. *Darwin's Black Box.* New York: Free Press, 1996.
[10] Brenner, S. as quoted in [52], 598.
[11] Brookbank. *Developmental Biology.* New York: Harper and Rowe, 1978.
[12] Chargaff, E. "Chemical Specificity of Nucleic Acids and Mechanisms of Enzymatic Degradation," *Experientia.* 1950, **6**, 201.
[13] Close, F. "The Quark Structure of Matter," *The New Physics,* Chapter 14 in [19].
[14] Crick, F.H.C. "On Protein Synthesis," *Symp. Soc. Exp. Biol.* 1958, **12**, 138
[15] Crick, F.H.C. "The Structure of the Hereditary Material, in [40], 65.
[16] Crick, F.H.C. "The Complementary Structure of DNA," *Proc. Natl. Acad. Sci.,* USA, 1954, **40**, 756.
[17] Crick, F.H.C. "General Nature of the Genetic Code for Proteins," *Nature,* 1961, **192**, 1227.

[18] Crick, F.H.C. "The Present Position of the Coding Problem," *Structure and Function of Genetic Elements: Brookhaven Symposium in Biology.* 1959, **12**.
[19] Davies, P. Editor, *The New Physics.* Cambridge: Cambridge University Press, 1995.
[20] Davies, P. *The Fifth Miracle.* New York: Simon and Schuster, 1999.
[21] Dawkins, R. *The Blind Watchmaker.* New York: W.W. Norton, 1986.
[22] Dayhoff, M.O., Eck, R.V. and Park, C.M. *Atlas of Protein Sequences and Structure,* **5**, Georgetown University Medical Center, Natl. Bioch. Research Foundation, 1972, 89.
[23] Dickerson, R.E. "Sequence and Structural Homologies in Bacterial and Mammalian-Type Cytochromes," *J. Molec. Biol.*, 1971, **57**, 1.
[24] Dyson, F. *Origins of Life.* Cambridge: Cambridge University Press, 1985.
[25] Eigen, M. and Schuster, P. *The Hypercycle: The Principle of Natural Self-Organization.* Berlin: Springer-Verlag, 1979.
[26] Franklin, R.E. and Gosling, R.G. "Evidence for Two-Chain Helix in Crystalline Structure of Sodium Deoxyribonucleate," *Nature,* 1953, **172**, 156.
[27] Franklin, R.E. and Gosling, R.G. "The Structure of Sodium Nucleate Fibres: I. The Influence of Water Content," *Acta Crystallographia,* 1953, **6**, 673.
[28] Franklin, R.E. and Gosling, R.G. "The Structure of Sodium Nucleate Fibres: III. The Three Dimensional Patterson Function," *Acta Crystallographia,* 1955, **8**, 151.
[29] French S. and Robinson, B. "What is Conservative Substitution?" *J. Molec. Evol.*, 1983, **19**, 171.
[30] Friefelder, D. *The DNA Molecule, Original Papers, Analysis and Problems.* San Francisco: Freeman, 1978.
[31] Gamow, G. "Possible Relation Between Deoxyribonucleic Acid and Protein Structure," *Nature,* 1954, **173**, 318.
[32] Garrod, Sir A.E. "A Contribution to the Study of Alkaptonuria," *Medico-Chirurgical Trans.,* 1899, **82**, 367.
[33] Gatlin, L.L. *Information Theory and the Living System.* New York: Columbia University Press, 1972.
[34] Gleick, J. *Chaos: Making a New Science.* New York: Penguin Books, 1988.
[35] Goodwin, B.C. and Trainor, L.E.H. "Tip and Whorl Morphogenesis in Acetabularia by Calcium Regulated Strain Fields," *J. Theo. Biol.*, 1985, **117**, 179.
[36] Grantham, R., Gautier, C., Gouy, M., Jacobzone, M. and Mercier, R. "Codon Catalog Usage as a Genome Strategy Modulated for Gene Expressivity," *Nucleic Acid. Res.* 1981, **9**, 43
[37] Griffiths, F. "The Significance of Pneumococcal Types," *J. Hygiene,* 1928, **27**, 113

References

[38] Grosjean, H., Sandorff, D., Minja, W., Fiers, W. and Gedergren, R.J. "Bacteriophage MS2 RNA: A Correlation between the Stability of the Codon: Anticodon Interaction and the Choice of Code Words," *J. Molec. Evol.*, 1978, **12**, 113.

[39] Haseltine,W.A. "Discovering Genes for New Medicine," *Special Report, Scientific American*. San Francisco: March 1997, 92.

[40] Haynes, R. and Hannawalt, P. "The Molecular Basis of Life," *American Scientific Readings*. San Francisco: W.H. Freeman, 1968.

[41] Hershey, A.D. and Chase, M. "Independent Functions of Viral Proteins and Nucleic Acid in Growth of Bacteriophage," *J. Gen. Physiol.* 1952, **36**, 39.

[42] Ho, Mae Wan *The Rainbow and the Worm: The Physics of Organisms*. Singapore: World Scientific, 1993 and 1998 (second edition).

[43] Ho, M.W. and Saunders, P. *Beyond Darwinism*. London: Academic Press, 1984.

[44] Holley, R.W. "The Nucleotide Sequence of a Nucleic Acid" in [40], 72.

[45] Hoyle, F. *The Intelligent Universe*. London: Michael Joseph, 1983, 19.

[46] Hubbard, R. and Wald, E. *Exploding the Gene Myth*. Boston: Bacon Press, 1993.

[47] Ingram, "A Specific Chemical Difference between the Globins of Normal and Sickle-Cell Anemia Haemoglobin," *Nature,* 1956, **178**, 792.

[48] Jacob, F. and Monod, J. *Proc. of the Fifth International Congress on Biochemistry, 1: Biological Structure and Function at the Molecular Level*, V.A. Englehardt (ed.), New York: Pergamon Press Inc., 1963, 132.

[49] Jaroff, L. "Keys to the Kingdom," *Time Magazine*, Special Issue, winter 1997-98, 56.

[50] Jian Min Yuan, *2000 S14 Mini Symposium on Complex Dynamics of Proteins*, SIAM Pacific Rim Dynamical Systems Conference, Maui.

[51] Jones, O.W. and Nirenberg, M.W. "Qualitative Study of RNA Code Words," *Proc. Natl. Acad. Sci.,USA*, 1962, **48**, 2115.

[52] Judson, H.F. *The Eighth Day of Creation*. New York: Cold Spring Harbor Press (Plainsview), 1996.

[53] Kauffman, S. *At Home in the Universe*. Oxford: Oxford University Press, 1995.

[54] Khorana, H.G. *The Harvey Lectures*, 1968, **62**, 79.

[55] Kornberg, A. *DNA Replication*. San Francisco: W.H. Freeman and Co., 1980.

[56] Kruskal, J.B. "Multidimensional Scaling by Optimizing Goodness of Fit to a Nonmetric Hypothesis,"*Psychometrica*, 1964, **29**, 1.

[57] Kruskal, J.B. "Nonmetric Multidimensional Scaling: A Numerical Method," *Psychometrica*, 1964, **29**, 115.

[58] Lewontin, R.C. "The Dream of the Human Genome,"*New York Review,* 1992, May 28 edition, 31.

[59] Lumsden, C.J. "Holism and Reduction," in [60], pp. 17-44.

[60] Lumsden, C.J., Brandts, W. A. and Trainor, L.E.H. (editors), *Physical Theory in Biology: Foundations and Explanations*. Singapore: World Scientific, 1997.
[61] Matthaei, J.H., Jones, O.W. Martin, R.G. and Nirenberg, M.W. "Characteristics and Compositions of RNA Coding Units," *Proc. Natl. Acad. Sci.*, USA, 1962, **48**, 666.
[62] Mendel, G. "Versuche uber Pflanzenhybriden" in Ostwald's *Klassiker der exacten Wissenschaften*, Leipzig, 1940.
[63] Meselson, M and Stahl, F.W. "The Replication of Escherichia Coli.," *Proc. Natl. Acad. Sci.*, USA, 1958, **44**, 671.
[64] Miller, R.V. "Bacterial Swapping in Nature," *Scientific American,* San Francisco: W.H. Freeman, January 1988, 66.
[65] Nirenberg, M.W. "The Genetic Code," in [40], II, 206.
[66] Nirenberg, M.W., Matthaei, J.H., and Jones, O.W. "An Intermediate in the Biosynthesis of Polyphenylalanine Directed by Synthetic Template RNA," *Pro. Natl. Acd. Sci.* USA, 1962, **48**, 104.
[67] Ochoa, S., from remarks of Crick and Nirenberg, 1965, in [40], pp.198, 205, 211, 222.
[68] Olby, R. *The Path to the Double Helix.* London: The Macmillan Press, 1974.
[69] Prigogine, I. *Non-Equilibrium Statistical Mechanics.* New York: Wiley-Interscience, 1963.
[70] Prigogine, I and Stenger, I. *Order out of Chaos.* Chapter 5, London: Heinemznn, 1984.
[71] Rein, R. *Perspectives in Quantum Chemistry, Intermolecular Interactions: From Diatomics to Biopolymers.* Paullman, B. (ed.). New York: Wiley, 1978.
[72] Rich, A. and Watson, J.D. "Some Relations between DNA and RNA," *Proc. Natl. Acad. Sci.*, USA, 1964, **40**, 759.
[73] Rifken, J. *The Biotech Century.* New York: Tarcher/Putnam, 1998.
[74] Rowe, G. W. "Information Content, Thermodynamics, and Codon Bias in Viral DNA," PhD Thesis, Department of Physics, University of Toronto, 1981.
[75] Rowe, G.W. and Trainor, L.E.H. "On the Informational Content of DNA," *J. Theor. Biol.* **101**, 151.
[76] Rowe, G.W. and Trainor, L.E.H. "A Thermodynamic Theory of Codon Bias in Viral Genes," *J. Theor. Biol.* **101**, 171.
[77] Rowe, G.W. Szabo, V. and Trainor, L.E.H. "Cluster Analysis of Genes in Codon Space," *J. Molecular Evolution*, **20**, 164.
[78] Rowe, G. W. "A Three-Dimensional Representation for Base Composition of Protein-Coding DNA Sequences," *J. Theor. Biol.,* 1985, **112**, 433.
[79] Sanger, F. "The Amino-Acid Sequence in the Phenylalanyl Chain of Insulin. 1 & 2," *Biochemical Journal*, 1951, **49**, pp. 463-481.

[80] Sanger, F and Thompson, E.O.P. "The Amino-acid Sequence of the Glycyl Chain of Insulin," *Biochemical Journal*, 1953, **53**, pp353-366.
[81] Schroedinger, E. *What is Life?* Cambridge: Cambridge University Press, 1944.
[82] Silver, L. *Remaking Eden; Cloning and Beyond in a Brave New World.* New York: Avon, 1997.
[83] Smith, M. "The First Complete Nucleotide Sequence of an Organism's DNA," *American Scientist*, 1979, **67**, 57.
[84] Snow, C.P. *The Two Cultures* (1959) and *A Second Look* (1963). Cambridge: Cambridge University Press.
[85] Spiegelman, S. "An In Vitro analysis of a Replicating Molecule," *Scientific American*, 1970, **210**(5), 48.
[86] Stent, G. "The Multiplication of Bacterial Viruses," 1968, in [40], 116.
[87] Stryer, L. *Biochemistry*. San Francisco: W.H. Freeman and Company, 1981.
[88] Trainor, L.E. H. "Emergence in Physics and Biology," 1987, in [60], 11.
[89] Trainor, L.E.H., Rowe, G.W. and Szabo, V.I. "A Tetrahedral Representation of Poly-Codon Sequences and a Possible Origin of Codon Degeneracy," *J. Theor. Biol.*, **108**, 459.
[90] Trainor, L.E.H., Rowe, G. and Nelson, G.J. "Codon Space: Exploring the Origins and Evolution of the Genetic Code," in [60].
[91] Trainor, L.E.H. and Wise, M.B. *From Physical Concept to Mathematical Structure: An Introduction to Theoretical Physics.* Toronto: Toronto University Press, 1979.
[92] Volkenstein, V. *Molecular Biophysics*. London: Academic Press, 1977.
[93] Wald, G. "The Origin of Life," [40], Chapter 34.
[94] Watson, J.D. *The Double Helix*, New York: Mentor, Penguin Books, 1969.
[95] Watson, J.D. and Crick, F.H.C. "Molecular Structure of Nucleic Acids: A Structure for Deoxyribose Nucleic Acid," *Nature*, 1953, **171**, 737.
[96] Watson, J.D. and Crick, F.H.C. "Genetical Implications of the Structure of Deoxyribose Nucleic Acid," *Nature*, 1953, **171**, 964.
[97] Watson, J.D. and Crick, F.H.C. "The Structure of DNA," *Cold Spring Harbor Symposium on Quantum Biology,* 1953, **18**,123.
[98] Wilkins, M.H.F. "Molecular Configuration of Nucleic Acids," *Science 140*. 1963, **3570**, 941.
[99] Wilson, E.O. *The Diversity of Life*. New York: W.W. Norton & Co., 1999.
[100] Wilson, E.O. *Consilience—The Unity of Knowledge.* New York: Vintage Books, Random House, 1999.
[101] Wilson, H.R. *Diffraction of X-Rays by Proteins, Nucleic Acids and Viruses.* London: Edward Arnold Publishers Ltd., 1964.
[102] Woese, C. "Evolution of the Genetic Code," *Naturwissenschaften,* 1973, **60**, 447.

Index

acupuncture, 75
adaption, 81, 85
adaptor molecules, 21, 23, 34, 94
adenine, 16, 18, 65, 93
adenosine triphosphate, 42, 93, 98
alanine, 41, 54
alkaptonuria, 32, 114
allele, 97
alpha helix, 26, 95, 103
amino acid, 2, 3, 4, 9, 12, 19, 20, 21, 23, 24, 25, 26, 27, 33, 34, 36, 37, 38, 39, 40, 41, 48, 52, 53, 54, 60, 62, 66, 67, 68, 69, 70, 72, 89, 90, 93, 94, 95, 98, 99, 101, 102, 103
aminoacyl-tRNA, 23
anabolism, 9
antibodies, 3, 72, 102
anticodon, 21, 22, 23, 24, 37, 54, 57, 94
archaea, 49, 99
arginine, 37, 40, 41
asparagine, 23, 37, 67
aspartic acid, 37
atomic nuclei, 28
ATP, 42, 93, 98
autocatalytic, 47
bases, 14, 15, 16, 17, 18, 19, 21, 22, 24, 32, 34, 35, 39, 40, 52, 53, 57, 65, 66, 90, 93, 94, 98
beta pleated sheets, 26, 95
biological information, 46
bosons, 72
carbon ring, 15, 16
carboxyl group, 25, 89, 101
catabolism, 9

chaos, 5, 114, 116
chaotic systems, 76
Chargaff's rules, 17
chicken and egg problem, 46, 47, 48
chromosomes, 3, 31, 33, 73, 83, 88, 97, 98, 109
citric acid cycle, 98, 99
cluster analysis, 56, 105, 113, 116
coherence, 5, 7, 74
common ancestor, 42, 52
complexity, 4, 7, 27, 28, 46, 79
complexity theory, 5
condon
    bias, 51, 53, 54, 55, 57, 66, 70
    families, 63, 66, 67, 70
    space, 51, 52, 57, 58, 59, 60, 61, 62, 105
conformation, 4, 17, 27, 85, 89
consciousness, 76
consilience, 78, 117
criticality, 76, 113
crossover, 35, 36
culture, 78, 85, 117
cystene, 37
cytosine, 16, 17, 18, 66, 93
Darwin, 45, 73, 88, 113
deBroglie, 97
defective genes, 111
degeneracy, 35, 36, 39, 40, 51, 52, 57, 66, 98, 117
denaturation, 27
deoxyribose, 13, 14, 16, 19, 93, 117
designer children, 83
development, xii, 1, 3, 4, 12, 97, 109, 111
diamond code, 40

119

differentiated, 33
differentiation, 8, 12, 91, 109
diploid system, 33
diversity, 2, 3, 7, 117
DNA, viii–x, 2, 4, 7, 8, 11, 12, 13, 15, 16, 17, 18, 19, 20, 21, 25, 26, 28, 31, 32, 33, 34, 35, 36, 37, 39, 40, 42, 47, 52, 55, 57, 58, 59, 60, 62, 68, 69, 72, 83, 88, 90, 92, 94, 96, 99, 105, 109, 110, 111, 112, 113, 114, 115, 116
DNA
   polymerase, 19, 40, 94
   replication, 19, 32, 42, 47, 70, 94
   structure, 11, 16, 32, 56, 93
DNA duplication, 96
DNA polymerization process, 56
drug design, 82
dynamic photography, 33
dynamical systems, 79, 102, 115
E. coli, 35
electrical laws, 28
electromagnetic fields, 1
electromagnetic waves, 92
electromagnetism, 87
electron microscope, 33, 97
emergence, xi, 73, 75, 117
endothermic, 9, 98
entropy, 8, 9, 46, 91
enzyme, 3, 4, 12, 19, 20, 21, 23, 28, 31, 32, 46, 47, 55, 72, 90, 92, 94, 98, 99, 101, 102, 110, 112
equilibrium, 9, 46, 47, 99, 116
escherichia coli (E. coli), 18, 116
eukaryote, 33, 51, 61, 94, 96, 109
evolution, xi, 3, 40, 42, 46, 47, 48, 55, 59, 72, 88, 102
exchange forces, 28, 96
exons, 21, 99, 112
factorial functions, 105
Feigenbaum numbers, 76
fermions, 72
fractal dimensions, 76

Genbank Library, 105
gene, 12, 13, 20, 30, 31, 32, 33, 35, 52, 57, 58, 61, 73, 81, 82, 83, 84, 92, 96, 97, 105, 109, 110, 111, 112, 115, 116
gene transplant, 81, 82
genetic
   code, viii–xii, 1, 2, 3, 4, 8, 18, 20, 27, 30, 34, 38, 41, 45, 46, 48, 51, 52, 57, 64, 66, 94, 98, 110, 113, 116, 117
   determinism, 5
   diseases, 83
   engineering, x, 77, 78, 79, 81, 83, 84, 85, 109, 110, 112
   genome, ix, 23, 33, 46, 52, 53, 56, 57, 81, 82, 83, 84, 97, 100, 109, 110, 112, 115
germ line, 82, 109, 111
glutamic acid, 37
glutamine, 37
God, 85
Golgi apparatus, 94
guanine, 16, 17, 18, 53, 66, 93
haploid, 33
hemoglobin, 103
histidine, 37
holism, 71, 72, 75, 76, 115
homo sapiens, 80
homologous chromosomes, 97
hormonal system, 25
hormones, 4, 72, 91
human disease, 77
human genome project, ix, 109
humanities, 78
hydrogen bonds, 15, 16, 17, 19, 55, 93, 95
hydrophilic, 25, 53, 90, 101, 102
hydrophobic, 25, 53, 90, 101
immune system, 25, 29
information, 2, 4, 11, 12, 15, 16, 17, 18, 20, 21, 30, 32, 33, 34, 35, 40, 45, 46, 51, 52, 58, 60, 62, 90, 95, 100, 109, 110, 114, 116

Index

inheritance, viii, 3, 4, 8, 11, 13, 31, 32, 33, 45, 72, 96
initiation factors, 20
inosine, 22
insulin, 91, 103, 117
introns, 21, 99, 112
inverse square law, 28, 96
irreversible changes, 91
Ising model, 55
isoleucine, 37, 41, 67
Kreb's cycle, 98
lac operon, 12, 92
leucine, 37, 40, 41, 67
levels of explanation, 76
life expectancy, 77
ligase, 110
lipids, 7, 90
lock and key, 28, 29
lysine, 37, 67
lysozome, 31
macromolecules, 7, 8, 25
MDS, 60, 61
meiosis, 33, 97
metabolites, 9, 80
methionine, 37, 69
Microtubules, 95
mitochondria, 41, 42, 61, 98
mitosis, 33, 96
molecular biology, xi, xii, 2, 5, 8, 12, 19, 34, 38, 40, 72, 73, 74, 78, 113
molecular Darwinism, 55, 100
mRNA, 4, 12, 20, 21, 22, 23, 24, 33, 36, 37, 38, 39, 40, 54, 57, 90, 94, 112
multi-dimensional scaling, 60
mutants, 31, 35, 97
mutations
　point, 36, 41, 70
natural sciences, 78, 79
neo-Darwinism, 73
nervous system, 25, 29
neurospora crassa, 31
neurotransmitters, 3, 29, 72
non-functional gene, 35
nonlinear
　dynamics, 5
　studies, 76
nonsense signals, 36
nucleases, 110
nucleoproteins, 31
nucleosides, 15
nucleotides, 16, 19, 36, 39, 53, 58, 94
organelles, 21, 41, 90, 94, 98
organisms, ix, xi, 2, 4, 7, 8, 18, 25, 33, 36, 41, 46, 47, 49, 51, 52, 56, 58, 61, 72, 75, 77, 79, 81, 85, 89, 93, 97, 99, 102, 109, 110, 115
organization, xi, 8, 9, 46, 77, 81, 84
origin of life, 45, 47, 49, 51, 52, 96, 99, 117
origins of life, viii, 51, 52, 62
overlapping code, 40
pairing, 54
pH, 25, 95
phage T4, 35
phages, 33, 61, 70, 110
phenylalanine, 37, 67
phosphate bonds, 93, 98, 99
plasmids, 110
polymerase
　RNA, 20, 90
polynucleotides, 48
polypeptides, 26, 27, 48, 89, 101, 102
polysaccharides, 7, 91
prebiotic, 4, 51, 55, 70
prokaryotes, 51
proline, 23, 37, 41, 67
promoter, 21
protein
　folding, 25, 27, 102, 103
　structure, 11, 25, 28, 33, 48, 93, 94, 95, 103, 109
proteomics, 85
protons, 28, 71
purine, 15, 16, 17, 93
pyruvate, 98, 99

quantization, 88
quantum
    mechanics, 1, 28, 72
    statistics, 5
quark confinement, 71, 74
quarks, 71, 73
receptor sites, 4, 28
recombinant DNA technology, 110, 111
recombination, 35, 98
reductionism, 71, 72, 73, 75, 76
regulation, 12, 46, 92
regulatory genes, 12
relativity, 74
    general, 74
    special, 74
replication, 19, 32, 42, 47, 70
residues, 18, 25, 26, 101
revelations, 1, 2
reversible changes, 91
ribose, 16, 19, 20
ribosome, 20, 21, 23, 24, 37, 90, 95
RNA
    messenger, 4, 20, 21, 33, 36, 90
    ribosomal, 21, 57, 90, 105
    transfer, 21, 22, 23, 34, 37, 54, 57, 105
rRNA, 21, 90, 94
self-organization, xi, 9, 46
serine, 37, 40, 53, 67, 98
specificity, 4, 23, 27, 30, 54, 90, 102
stability, xii, 8, 49, 54, 55, 70, 81, 115
stacking energies, 55, 56

statistical mechanics, 70
strings, 2, 3, 9, 12, 14, 25, 27, 28, 72, 95
substrates, 28, 29
superbugs, 49, 99
superconductivity, 76
superfluidity, 76
tetrahedral representation, 51, 62
tetrahedron, 62, 63, 64, 65, 66, 67, 68, 69, 107
thalidomide, 82
thermodynamics, 8, 52, 55, 91
threonine, 37, 41, 67
thymine, 16, 18, 19, 93
tiling problem, 107
transcribed, 20, 21, 33, 92, 112
transcription, 4, 12, 20, 37, 39, 40, 49, 70, 81, 92, 112
transgenetic, 81
triplet code, 4, 36, 39, 53, 62, 68
tRNA, 21, 22, 23, 24, 34, 54, 94
tryptophan, 37
tyrosine, 37, 67
universality (of genetic code), 11, 37, 41, 46, 49, 52, 110
valine, 37, 41
vitalism, 7
Watson-Crick base pairing, 54
Watson-Crick pairing, 17, 21, 55, 90, 93, 94
X-ray analysis, 3, 11
zygote, 91, 97